David Syme

Modification of organisms

David Syme

Modification of organisms

ISBN/EAN: 9783742865007

Manufactured in Europe, USA, Canada, Australia, Japa

Cover: Foto ©berggeist007 / pixelio.de

Manufactured and distributed by brebook publishing software
(www.brebook.com)

David Syme

Modification of organisms

ON THE

MODIFICATION OF ORGANISMS

BY

DAVID SYME

MELBOURNE: GEORGE ROBERTSON AND CO.
LONDON: KEGAN PAUL, TRENCH AND CO.

CONTENTS.

CHAPTER I.

THE PROVINCE OF NATURAL SELECTION.

PAGE

What natural selection is not—What it is—Darwin's explanation unsatisfactory—Selection by nature and by man—The province of natural selection—Variation a preliminary to the operation of natural selection—A fatal admission—Darwin's definition misleading—His statements irreconcilable—Advocates two kinds of natural selection—Refutes his own theory—Summary   1

CHAPTER II.

THE EFFECTS OF NATURAL SELECTION.

Natural selection not preservative—The case of protective coloration—The facts admitted—The alleged cause disputed—Action of light and heat in producing colour—Only the more exposed parts of animals protectively coloured—Voluntary protective coloration—The analogy of vegetable coloration—A mechanical process not needed—The case of instinct—Darwin's definition of—His theory of the origin of unsatisfactory—Dr. Romanes on the incubating instinct—Primary and secondary instincts—Theories of Herbert Spencer and Lewes—The nature of instinct—Contrasted with intelligence—Summary   17

CHAPTER III.

THE EXTERMINATION OF THE UNFIT.

PAGE

The survival of the fittest—Natural selection demands the extermination of intermediate forms—Absence of evidence—The case of the *foraminifera* proves the contrary —Inferior races not usually exterminated by the superior— Views of Malthus—Case of the human species—Effects of the struggle for existence—Tendency of natural selection —Herbert Spencer on the effects of competition—An illustrative case—Causes which bring about the extinction of species—Summary 42

CHAPTER IV.

SEXUAL SELECTION.

Darwin's definitions of—His views on the transfer of secondary sexual characters—Size and strength—Weapons of offence —Colour, sound, and odour—Other special secondary sexual characters—For what purpose acquired—Darwin's views—Polygamous eared seals—Primitive marriage— Selection of the male by the female—Want of harmony between natural selection and sexual selection—Summary 59

CHAPTER V.

THE FERTILIZATION OF PLANTS BY INSECTS.

Relationship between plants and insects—Floral attractions— The nectar and the nectary guides—The *Coryanthes*— The argument from structure—Dichogamy—The law of Use and Disuse—Proterandry and proterogyny—The effects of cross-fertilization—Darwin's experiments—How vitiated—*Ipomea purpurea*—*Mimulus luteus*—*Eschscholtzia californica*—*Petunia violacea*—Darwin's conclusions— The distribution of seeds—Vigour and fertility of plants from self-fertilized seed—Cause of uniformity in nature— Comparative scarcity of entomophilous flowers—Occa-

PAGE
sional failure of insect fertilization—Inconspicuousness of tropical flowers — Characteristics of alpine flora—Injurious effects of insect visitations—What insects are supposed to be useful—Action of humble bees—Of hive-bees—The beauty of flowers—Summary 75

CHAPTER VI.

THE CAUSES OF VARIABILITY.

Action of external conditions on the organism—Schmankewitsch's experiments with *Artimia*—Semper's experiments with *Lymnœa stagnalis*—Influence of situation—Effects of intercrossing—Climate—Darwin and Wallace on the origin of variations—The action of the organism—*Larus argentalus*—Organic changes due to physiological causes—Analogy of Language—Summary .. 118

CHAPTER VII.

THE MODIFYING AGENCY.

Difference between a law and a cause—The intelligence of vegetative organisms—Movements in fertilization—*Trapa natans*—Intelligence in the lower forms of animals—The Rizopod—The Amœba—The Arcella—Intelligence of the cell—in the construction of the organism—in expelling foreign bodies—in restoring lost organs—Inadequacy of reflex action to account for the phenomena—The cell the source of variations—Lamarck's theory—The cell the biological and psychological unit—On the term *Mind*—Mental phenomena not the result of organization—Life and Mind—Unity and diversity—The Ego and the Soul—Development of the Ego—Summary 131

APPENDIX.

A.—MIMICRY 158
B.—THE CELL 160
C.—PERSONAL IDENTITY .. 162

CHAPTER I.

THE PROVINCE OF NATURAL SELECTION.

AFTER a review of all that has been said for and against Natural Selection during the last thirty years, Dr. Wallace expresses himself, in his recent work on Darwinism, as having come to the conclusion that this principle " is supreme to an extent that even Darwin himself hesitated to claim for it." * Notwithstanding the opinion of this high authority, I venture to think there is still something to be said on the other side of the question, and I propose in the following pages to show that the acceptance of this theory is still beset with difficulties of a very formidable character.

The central idea of Darwin's system is natural selection, and it will therefore be necessary to start with a clear perception of what natural selection is, and what it is not. The necessity of this course will be the more

* *Darwinism*, p. 444.

2

apparent as we proceed in our inquiry; for we shall find that Darwin himself did not always bear in mind his own definition, and often allowed himself a latitude of expression which led him to erroneous conclusions. Although he has nowhere given us a precise definition of the term natural selection, he has in various places indicated, more or less plainly, what he means by it. He tells us, in the first place, that it is not a creative process, that it is not the cause of organic variability, but that, as the term itself implies, it is a purely selective process. Natural selection only preserves variations already provided for it. Variations are "given by the hand of nature." *   "Natural selection," he says, "depends on the survival under various complex circumstances of the best fitted individuals, but has no relation whatever to the primary cause of any modification of structure." †   Indeed, he distinctly and emphatically repudiates the idea that natural selection is the cause of variability.   "Some have imagined," he says, "that natural selection induces variability, whereas it implies only the preservation of such variations as arise." ‡   And again, "Natural selection acts *only* by the preservation and accumulation of small inherited modifications, each profitable to the preserved being." §   These statements

---

* *Origin of Species*, 7th ed., p. 49.
† *Variation of Animals and Plants*, vol. ii., p. 272.
‡ *Origin of Species*, p. 63.
§ *Variation of Animals and Plants*, vol. ii., p. 75; *see also* p. 97.

are clear enough, and it will be well to bear them in mind, for Darwin, strange to say, appears to have forgotten them after putting them on record, and, in the course of his work, announces a theory altogether inconsistent with them.

Having cleared the way so far by showing, from Darwin's own words, what natural selection is not, we shall now proceed to state, again from Darwin's own words, what natural selection is. "The preservation of favourable individual differences and variations, and the destruction of those which are injurious, I have," he says, "called natural selection."* Natural selection is therefore another name for the struggle for existence. I cannot help thinking that the latter is much the better expression of the two, being less ambiguous. For the same reason it is preferable to another expression which Darwin seems to regard as equivalent. "The principle of preservation, or the survival of the fittest," he says, "I have called natural selection."† But the two expressions are not exactly equivalent. Selection is a process; survival of the fittest is not so much a process as the result of a process. Here Darwin has inadvertently shown on what a grave error his system rests. He has put the effect for the cause. Variation is the cause, and selection, or the survival of the fittest, is the effect. Natural selection, then, according to Darwin, operates by preserving the useful and by destroying the useless varia-

* *Origin of Species*, p. 63.　　† *Ibid.*, p. 103.

tions, and it does this by leaving the useful and the useless to struggle for existence as best they may.

It will be readily admitted that on one point Darwin's explanation is ambiguous and unsatisfactory. The point happens to be most essential, for it refers to the province or function of natural selection. Having assured us that natural selection only preserves and accumulates useful variations, he subsequently qualifies this statement by another which is completely at variance with it. Thus he tells us that "when a variation is of the slightest use to any being, we cannot tell how much to attribute to the accumulative action of Natural Selection and how much to the definite actions of the conditions of life."* Nothing, by the way, is here said about the influence of the organism. According to this a useful variation is, therefore, the result either of natural selection, or of the conditions of life. In that case it would be the antecedent, rather than the result, of natural selection, and thus it is made to appear that natural selection is not a selective process merely, but also a creative process, which is in direct contradiction to the principle already laid down. We shall find more of this sort of ambiguity as we proceed, for it runs through the whole of Darwin's exposition.

We have now to determine what is the actual province or function of this mysterious agent which Darwin

* *Origin of Species*, p. 107.

invests with such elastic powers. At first sight it might appear as if there were some analogy between selection by man and selection by nature. In both cases the variations are provided; and, in a certain sense, the struggle for existence may be regarded as selective, since the fit are preserved and the unfit destroyed. But here the analogy ends. The breeder starts with an ideal form, and he selects the best animals for the purpose he has in view; nature, on the other hand, has no ideal, and makes no choice. Having selected his particular type of animal, the breeder takes means to prevent his females from interbreeding with males of a different type, and his males from herding together or fighting with one another, and he is careful to provide his stock in all seasons with a plentiful supply of food. Nature, on the other hand, puts no restriction on intercrossing, takes no steps to prevent the males from fighting and injuring one another, and places no check on the increase beyond the means of subsistence. Darwin describes the action of natural selection as preservative and accumulative, but, properly speaking, it is a purely destructive process. It is heredity, and not natural selection, which is preservative and accumulative.

Darwin's position, then, is this: "Natural selection can do nothing unless favourable individual differences arise." * Individual differences, however, are of two kinds, the useful and the non-useful, the favourable and

* *Origin of Species*, p. 137.

the non-favourable, for if there were only one kind there would be no material for natural selection to operate on. Having got the variations, therefore, both the favourable and the non-favourable, natural selection is supposed to step in, and do—what? Nothing, apparently. Everything has already been done that requires doing. The variations have been provided, the favourable as well as the non-favourable; the superiority of the former has already been proved by the fact that they have been adopted by the organism, as otherwise we should not have known them to be favourable. Natural selection can only consist in the selection of the favourable variations, and these, as we have said, have been proved to be such before natural selection appears on the scene. It is not an independent force whose function it is to make a choice between two or more variations presented to it. Darwin calls it the struggle for existence; but it is the organism which struggles—not, however, to select this or that variation, but to adapt itself to its environment.

It would seem, indeed, as if Darwin had a suspicion that all was not well with natural selection, and that it was not so potent a factor in organic modification as he represented it to be, for in one place he ascribes to the conditions of life all the potency which he, in other places, claims for natural selection. " In one sense," he tells us, " the conditions of life may be said, not only to cause variability, either directly or indirectly, but likewise to include natural selection ; for the conditions

determine whether this or that variety shall survive." *
In another place he goes even farther, and makes the
following damning admission:—" It should not be overlooked that certain rather strongly marked variations,
which no one would rank as mere individual differences,
frequently occur, owing to a similar organization being
similarly acted on. .. . . There can be little doubt
that the tendency to vary in the same manner has
often been so strong that all the individuals of
the same species have been similarly modified *without the aid of any form of natural selection.*" †
If " strongly marked variations " ("not mere individual differences," as he is candid enough to point
out), can be established without the aid of " any form "
of natural selection, of course less strongly marked
variations might also be established in the same way ;
and if " *all* the individuals " of one species can be modified in this manner, why may not all the individuals
of any or every species be similarly modified ?   Dr.
Wallace seems somewhat uncomfortable under this admission, and is of opinion that Darwin made a slip with
his pen when he wrote the word "all" in the statement we have quoted.‡

Had Darwin stopped here in his exposition, the principle of natural selection would not have appeared to

* *Origin of Species*, p. 107.   † *Ibid.*, p. 72.
‡ *Darwinism*, p. 141.

anyone as a very great discovery. It would have been looked upon merely as the old familiar theory of the Balance of Nature under a new name. Just at this point, however, Darwin makes a new departure. The modest claim he put forth at the outset on behalf of natural selection does not content him as he proceeds. Gradually, and it may be unconsciously, he extends the province of natural selection, and unfolds what is really an altogether new theory, and a theory, moreover, utterly inconsistent with the one he started with, and of which an outline has been given. Darwin never acquired the art of using precise language to convey his meaning, consequently one never knows whether the terms he makes use of are to be taken in their natural or in a figurative sense. He admits that the expression natural selection, as used by himself, is misleading—nay, that it is false.* Had he substituted for natural selection the expression "struggle for life," there would, it is true, have been less novelty about it, but there would also have been less liability to error, both on his own part and on the part of his readers. We have seen that he defines natural selection as "the struggle for existence," and again as "the survival of the fittest;" but while thus identifying the two expressions, we find him also distinguishing between them. In one place he endeavours to explain "how the struggle for existence

* In the literal sense of the word, no doubt, natural selection is a false term.—*Origin of Species*, p. 63.

bears on natural selection,"* and in another he speaks of "the struggle for existence *and* the principle of natural selection."† Such inaccuracies of expression occur in almost every page of his writings. Darwin is quite as loose in his statements as in his terms. Thus, after assuring his readers over and over again that natural selection only selects or preserves; that it is not the cause of variability, and cannot even induce it; that, in fact, it is absolutely powerless to effect anything until the variations are provided for it, he, nevertheless, proceeds, and without notice, to show that this same natural selection is a far-reaching and veritable transforming power, and capable of bringing about the very changes which he had, in express terms, disclaimed in advance on its behalf. Thus, in one place, he says:—"What applies to one animal will apply throughout all time to all animals, that is, if they vary, for otherwise natural selection can effect nothing."‡ While, in another place, he tells us that "it would be easy for natural selection to adapt the structure of the animal to its changed habits;"§ that he sees no difficulty in being able to prove that natural selection " is competent to account for the incipient changes of useful structures."‖ So far, indeed, does he go in this direction that we find him distinctly announcing that the production of protective

* *Origin of Species*, p. 48. † *Ibid.*, p. 143.
‡ *Ibid.*, p. 88. § *Ibid.*, p. 144.
‖ *Ibid.*, p. 198.

coloration, the formation of the eye, and even of such organs as are used only once in a lifetime, are all due to the operation of natural selection. Thus, he says, "Natural selection might be effective in giving the proper colour to each kind of grouse, and in keeping that colour, when once acquired, true and constant."* Speaking of the eye, he says :—"Although the belief that an organ so perfect as the eye could have been formed by natural selection is enough to stagger anyone, yet in the case of any organ, if we know of a long series of gradations in complexity, each good for its possessor, then, under changing conditions of life, there is no logical impossibility in the acquirement of any conceivable degree of perfection through natural selection."† Of organs used only once in a lifetime, such as the hard tip to the beaks of unhatched birds, used for breaking the egg, or the great jaws possessed by certain insects, used for opening the cocoon, he contends that "if of high importance to the animal," they "might be modified to any extent by natural selection."‡ In fact, he tells us that "*any* change in structure and function, which can be effected by small stages, is within the power of natural selection."§

It is evident that we have here two kinds of natural selection. We have a natural selection which selects or preserves only, and we have another which adapts,

* *Origin of Species*, p. 66.   † *Ibid.*, p. 165.
‡ *Ibid.*, p. 68.   § *Ibid.*, p. 401.

modifies, or creates. According to the one, variations are produced without the aid of natural selection; according to the other, they are the natural and direct result of that process. It is needless to say that if natural selection be merely selective, it cannot at the same time be transformative; and, if it be transformative, it would be absurd to call it merely selective. The two systems are fundamentally different. Which are we to accept? He nowhere informs us.

We may here point out that Darwin is more than inconsistent—he actually refutes himself. In his earlier work he attempts to prove that species originated " by means of natural selection," while in his *Plants and Animals under Domestication* he lays down certain principles which practically dispose of that theory. Thus, in the *Origin of Species*, he starts with two assumptions which he considers absolutely indispensable to his system. The first is that variations must be "slight," or small in amount. " Natural selection," he says, " acts only by taking advantage of slight successive variations; she can never take a sudden leap, but must advance by short and sure, though slow, steps." * " If it could be demonstrated that any complex organ existed which could not possibly have been formed by numerous successive slight modifications, my theory would absolutely break down." † This assumption has not been verified; and there are many facts which seem to prove

* *Origin of Species*, p. 156.   † *Ibid.*, p. 227.

that it cannot be established, as, for instance, the metamorphoses of the embryo, the changes in insect development, and the sudden appearance of new organs of which not a trace can be found in the embryo or in the adult forms lower in the scale. But this only by the way. The second assumption is that favourably modified individuals should be few in number, "two or more" * being considered sufficient. Now, if natural selection is not to be a merely nominal factor in the origination of species, this assumption is absolutely necessary to his theory, as it is obvious that if large numbers of individuals were spontaneously modified in the same manner natural selection would be forestalled.† In his *Plants and Animals under Domestication* Darwin completely refutes these assumptions. With regard to the first, he assures us that " it is obvious that free crossing," which would be inevitable under the circumstances, "would obliterate such small distinctions as those referred to;" ‡

---

* *Plants and Animals under Domestication*, vol. ii., p. 7.

† To show that I am not misinterpreting Darwin by this remark, I quote the following from *Animals and Plants under Domestication*, vol. ii., p. 279 :—" By the term definite action, I mean an action of such a nature that, when many individuals of the same variety are exposed during many generations to any change in their physical conditions of life, all, or nearly all the individuals are modified in the same manner. A new variety would thus be produced *without the aid of selection.*" Now, Darwin's object is to prove that natural selection, and not the conditions of life, is the main factor in the formation of species.

‡ *Ibid.*, vol. ii., p. 90; *see also* p. 173.

while as regards the second he disposes of it by the statement that " when one of two mingled races exceeds the other greatly in number, the latter will be wholly, or almost wholly, lost," * as in the previous instance, by sheer force of numbers.

Darwin's followers imagine they get over this difficulty by supposing that the modified individuals may become isolated from the others. But Darwin effectually blocks this way of escape. Let us suppose that a pair, consisting of one of the modified and one of the unmodified individuals, wandered away from the parent stock and bred by themselves ; the progeny, in this case, would not partake of the characteristics of each parent, for here the law of Prepotency of Type would come into force. The older a type the more prepotent it is, or the greater its tendency to transmit its character to its offspring ; hence the importance which breeders attach to a long pedigree. Darwin admits the force of this when he says that "it is obvious that a purely bred form of either sex . . . will transmit its character with prepotent force over a mongrelized or already variable form," † such, for instance, as a casual variation.

Would the result be any different if a pair of the modified forms were isolated from the rest and reared a progeny ? Not in the least ; for here again, they would

* *Plants and Animals under Domestication*, vol. ii.. p. 247.
† *Ibid.*, vol ii., p. 69.

be blocked by the law of Reversion. It is well known among breeders of domestic animals and plants that when the offspring of crossed parents breed with each other, their progeny will usually be unlike either parent, and if the intercrossing be continued that their descendants will have a tendency to revert to some archaic type. Darwin thoroughly endorses this opinion. " When two races or species are crossed," he says, " there is the strongest tendency to the reappearance in the offspring of long-lost characters, possessed by neither parent nor intermediate progenitor." * If this is true as regards races, it will also be true as regards variations, which are even more unstable than the former, " a new character, or some superiority in an old character," being, he says, " at first faintly pronounced," and "not strongly inherited." † But this is not all ; for if two individuals of the same race intercross with each other, and their progeny go on intercrossing, as in the last mentioned instance, . we should have interbreeding of the closest kind, and Darwin affirms that " interbreeding prolonged during many generations is highly injurious," resulting in loss of vigour, size, and fertility, and tending to malformations. ‡ So that, whether the two forms mingle, or whether they separate, the probability of any new variety or species being established by the process of

---

* *Plants and Animals under Domestication*, vol. ii., p. 48; *see* also p. 43. · ·

† *Ibid.*, vol. ii., p. 193.   ‡ *Ibid.*, vol. ii., p. 143, 144.

natural selection is, according to Darwin's own principles, reduced to a minimum.

Summary.—We have seen that Darwin's language is wanting in precision, and his definitions and theories are variable and contradictory. In one place natural selection is the "struggle for existence;" in another, the "struggle for existence" is said to "bear on" natural selection; in a third place he speaks of the "struggle for existence and natural selection" as if they were independent principles; in one place, again, he defines natural selection as "the survival of the fittest," thus confounding cause with effect, and in another place he says that natural selection "depends on" the survival of the fittest; while, to add to the confusion, he tells us in another place that "the conditions of life include natural selection," inasmuch as they determine whether this or that variety shall survive. In numerous places he explains that the function of natural selection is merely selective, as the term implies, that it operates on variations which are provided for it, and is absolutely powerless to effect anything without them ; in other places he insists that variations are created by natural selection, and that, in fact, every change in structure and function is within the power of natural selection. Unless we are to assume that variations are not changes in structure or in function, then these two theories are irreconcilable. We have seen that the principles laid down by Darwin with regard to breeding are not in accord with his theory of natural selection; and, finally, that Darwin has practically abandoned his theory altogether, when he admits that the tendency to vary in the same manner is so strong that whole species may be modified without the aid of any form of natural selection.

CHAPTER II.

WE have seen what Darwin means by natural selection
—that it is the struggle for existence, the striving of
the organism to maintain itself against adverse conditions of life, animate and inaniniate. The nature
of the struggle will of course depend on the nature of
the conditions. There is the struggle for life, and the
struggle for subsistence. Between the Carnivora and the
Herbivora the struggle is for life ; between organisms
of the same or of allied species, the struggle is for subsistence ; and, in addition to both of these, there is the
struggle against the inorganic forces of nature. Natural
selection is, in fact, a severe form of competition.

We have also seen that Darwin has put forth two
distinct and contradictory theories of the functions of
natural selection. According to the one theory natural
selection is selective or preservative, and nothing more.
According to the other theory natural selection creates
the variations, and we are left to infer that it afterwards
selects them, as we cannot conceive how it could select
at all before there were variations to select from. Of
these two views Darwin evidently favours the one which
affords the widest scope for the operation of natural

3

selection, and it would serve no good purpose for us to bind him to the other and narrower meaning with which he set out. We shall, therefore, set aside all Darwin's definitions and explanations which would limit the functions of natural selection, and we shall assume that it is a creative as well as a preservative and destructive process.

It certainly seems absurd to speak of natural selection, or the struggle for existence, as selective or preservative, for the struggle for existence does not preserve at all, not even the fit variations, as both the fit and the unfit struggle for existence, the unfit naturally more than the fit, and the fit are preserved, not in consequence of the struggle, but in consequence of their fitness. Suppose two varieties of the same species are driven, by an increase of their numbers, to seek for subsistence in a colder region than they have been accustomed to, and that one of these varieties had a hardier constitution than the other; and let us suppose that the former withstood the severe climate better than the latter, and consequently survived, while the other perished. In this case the hardier survived, not because of the struggle, but because it had a constitution better adapted to the climate. I wish to ascertain if a certain metal in my possession is gold or some baser metal, and I apply the usual test; but the mere fact of my testing this metal would not make it gold or any other kind of metal.

I venture to dissent altogether from Darwin on the

question of the functions and tendency of natural
selection. I maintain that natural selection does not
create the favourable variations, at all events in the sense
understood by him, and that it does not even preserve
them. I go further than this and assert that it does not
even exterminate the unfavourable variations. I shall
endeavour to show that it is neither creative, preservative, nor greatly destructive; that it neither produces nor
preserves the fit nor exterminates the unfit, and that so
far from being beneficent in its operation, as Darwin
and his followers represent, that the struggle for existence
is, on the whole, pernicious, and tends to produce disease,
premature decay, and general deterioration of all beings
subjected to its influence. In support of my contention,
I shall first refer to certain facts or cases which Darwinists rely on to prove the truth of their theory.

That protective colouring is an advantage to the
animal possessing it no one will dispute; that it has
been acquired by natural selection no Darwinist enter-
tains the smallest doubt, least of all Darwin himself.
"When we see leaf-eating insects green, and bark feeders
mottled grey, the alpine ptarmigan white in winter, the
red grouse the colour of heather, we must believe that
these tints are of service to those birds and insects in
preserving them from danger;" hence, he says, natural
selection is "effective in giving a proper colour to each
kind of grouse, and in keeping that colour, when once

acquired, true and constant."* According to Darwin, the process by which protective coloration has been brought about is simplicity itself. We have only to suppose the occurrence of an endless number of colour variations, and the fortunate possession by one or more individuals of colour of the right sort—that is, a colour similar to that of surrounding objects—and the destruction of all animals with non-protective colours will follow as a matter of course, as animals with non-protective colours, being conspicuous, will be picked off by their carnivorous enemies, while the protectively coloured will escape observation and survive. According to Dr. Wallace variations in colour are purely accidental. Speaking of imitative coloration, he says:—"To many persons it will seem impossible that such beautiful and detailed resemblances as those now described—and they are only samples of thousands that occur in all parts of the world—can have been brought about by the preservation of accidental useful variations. But this will not seem so surprising if we keep in mind the facts set forth in our earlier chapters, the rapid multiplication, the severe struggle for existence, and the constant variability of these and other organisms. And, further, we must remember," he goes on to say, "that these delicate adjustments are the result of a process which has been going on for millions of years, and that we now see the small percentage of successes among the myriads of

* *Plants and Animals under Domestication*, p. 66.

failures."* Darwin also speaks of variations as fortuitous or accidental, but he is careful to explain that he uses such terms provisionally, as indicating the unknown causes of phenomena. But whatever opinion one may entertain as to the cause, no one will dispute the fact that many animals are protectively coloured. The rabbit, the hare, the deer, are coloured like the cover which conceals them; and so are the lark, the quail, and the grouse among birds. Carnivora, like the lion, the tiger, and the panther have also protective colouring, and are thereby enabled to steal upon their prey unawares. Insects resemble the leaves or the bark of the trees or plants on which they rest; fishes take the colour of the ground they feed upon. Thus, flounders are of a motley brown colour, like the gravelly bottom they frequent; eels live in muddy water, and are mud coloured; fish which feed among rocks are coloured like the weeds which adhere to them. Individuals of the same species also acquire colours according to the nature of their environment for the time being. Thus, trout which frequent a clear stream with a gravelly bottom, are much lighter in colour than those which live in the same stream where the water is less transparent. Many animals, again, change their colour with the seasons—are grey, brown, or black in summer, and white in winter. The Arctic fox,

* *Darwinism,* p. 205. *See also* p. 244, where he speaks of "the accumulation of *fortuitous* variations" in the same connection.

hare, and ermine, for instance, as well as many alpine
animals, are dark-coloured in summer and white in
winter. Some British insects change their colour in
summer and autumn, so as to resemble the vegetative
tints peculiar to those seasons. Caterpillars also have
their seasonal changes, some being brown in autumn,
corresponding to the fading foliage, and bright green in
spring, like the foliage of that period of the year.*

It is possible to conceive that permanent colours may
be acquired by natural selection, but the theory is
inapplicable to alternative or seasonal colours. Natural
selection is supposed to render a variation stable, but a
variation cannot be rendered stable and yet remain
unstable; it cannot be permanent and at the same time
be alternating. The struggle for existence is supposed to

---

* Referring to the various species of Phasmidæ and Mantidæ,
which he met with in Africa, Professor Drummond says :—They
are variously coloured according to season and habitat. When
the grasses are tinged with autumn tints they are the same,
and the colours run through many shades, from the pure bright red,
such as tips the fins of a perch, to the deeper claret colours, or the
tawny gold of port. But an even more singular fact remains to
be noted. After the rainy reason, when the new grasses spring
up with their vivid colour, these withered grass insects seem all to
disappear. Their colour now would be no protection to them, and
their places are taken by others coloured as green as the new
grass. Whether these are new insects, or only the same in spring
toilets, I do not know, but I should think they are a different
population altogether, the cycle of the former generation being,
probably, complete with the end of summer.— *Tropical Africa*,
pp. 172-3.

exterminate the conspicuous animals, in this case the
unfit, and to preserve the non-conspicuous, in this case
the fit; but if all the brown-coloured animals, for
instance, were destroyed, and the white-coloured only
remained, the brown could not alternate with the white
in the same individual, as the brown has been wholly
eliminated; on the other hand, if all the white-coloured
animals were exterminated and the brown-coloured only
remained, the white could not alternate with the brown,
as the white has altogether disappeared. When a colour
has once been acquired by natural selection, it is established to the exclusion of all other colours for all time.
The very principle of natural selection renders alternative coloration in the same individual an impossibility.
If it be said that natural selection might preserve the
individuals having a tendency to assume alternative
colours under certain conditions, as, for instance, brown
in summer and white in winter, then I answer that the
change in colour would depend on the change in the
conditions, and not on natural selection. The conditions
would, therefore, be the predisposing cause, and natural
selection would be inoperative.

The acquisition of protective colouring is not to be
explained on mechanical principles. Coloration is an
organic or a physiological process, and must, therefore,
have an organic or physiological cause. But the Darwinist looks upon organic phenomena as he would upon
the motions of an ingenious piece of mechanism. He sees

the mechanism in motion; he observes the cranks turn, the cylinders move, the shaft revolve, and the big ship propelled through the water, and he calls upon us to admire the beauty of the whole arrangement. He never inquires about the motive power. Had he looked below the engine-room he would have seen, by the glare of the furnace fires, rows of stalwart, perspiring, half-naked men piling the fuel on the fires, which heat the boilers, which supply the motive power, which puts the whole machinery in motion.

It is highly probable that the coloration of animals and of plants is produced in a somewhat similar manner —namely, by the action of light, and, to a certain extent, of heat, on the pigment and chlorophyll cells. The colour of the human skin is black or dark brown under the equator, olive in the sub-tropical, and almost white in the temperate regions of the earth, while in the Arctic regions, where the sun is not visible during a great portion of the year, white is the predominant colour of the animals which live there during that period. Cavernous animals and ento-parasites, which never see the light, are almost invariably white. In this respect there is a close analogy between plants and animals. It has been demonstrated that the green colouring matter of plants cannot be produced without the co-operation of sunlight; and it is well known that the leaves of plants assume various shades of colour according to the amount of light to which they are exposed.

We have next to observe that it is only those parts of the bodies of animals which are most exposed to the action of light that are protectively coloured, the less exposed portions having, as a rule, no protective colouring whatever. Thus, the upper surface of rabbits, hares, and hawks is grey or brown, the under side is of a whitish hue. In the case of certain butterflies, on the contrary, it is the under side of the wings that is protected and not the upper side; but this exception rather proves the rule than otherwise, as butterflies, when at rest, put up their wings and expose the under side to the action of the sun. Thus it happens that butterflies have their brilliant non-protective tints on the upper surface of their wings, while the under side is almost invariably protectively coloured.

In some cases protective coloration is produced by the voluntary action of the animal. It is well known that certain fishes, as the stickleback, the perch, certain species of serranus and salmon, rapidly change their colour, and apparently at will, for the purpose of escaping observation. It has been ascertained that the pigment cells have the capacity of expansion and contraction, and that the colour of the animal varies according to the manner in which this power is exercised. The experiments of Pouchet and others show that the arrangement of the pigment cells in some manner depends on sight, as when the optic nerve is divided the activity of the cells ceases, and no corresponding change in colour takes place.

The phenomena of protective or permanent coloration cannot, however, be explained on this principle.

So far it is certain that light and heat are essential factors in the production of animal colouring, and we have now to ascertain how these operate. Here again the analogy of vegetable coloration will assist us. It is well known that when the chlorophyll cells are exposed to diffused light they arrange themselves in planes perpendicular to the direction of the incident ray (epistrophe), but when they are exposed to bright sunlight they form in planes parallel to the direction of the incident ray (apistrophe), with the result that in the former case the leaves assume a light, and in the latter case a dark green colour. That the action of light alone, or light and heat combined, is not the cause of protective coloration, is evident from the fact that wool, hair, and fur, when removed from the skin of a living animal, react upon these agencies in a different way, and become lighter in colour instead of darker when exposed to the action of the sun. A similar result is produced on dead vegetable matter; the most brilliantly coloured seaweed, for instance, becomes perfectly white when exposed for a few days on the seashore. The action of sunlight on the chlorophyll cells produces every variety of vegetative tint, and we are justified in concluding that the sun's rays, acting on the pigment cells, produce a corresponding effect on the skins of animals. In both cases, the process is very similar. The chlorophyll and the pigment cells

display the same kind of movements; they react under
similar conditions; and similar causes, acting under
similar conditions, should produce similar effects.
There are many facts which seem to favour this view.
Animals change their colour at the same periods of the
year that plants change their foliage; animals are
affected by the vicissitudes of the seasons as well as
plants, especially in spring-time and autumn; the
moulting of the feathers and the shedding of the hair,
wool, and fur, are concurrent with corresponding changes
in vegetation. Many Arctic animals change their coats
as many as four times during the year—are dark-
coloured in summer, white in winter, and an intermediate
colour in spring and autumn. Then, again, corresponding
with the seasonal alterations in vegetation, many animals
lose their protective colouring when in confinement,
owing, probably, to their being less exposed to the action
of the sun, as in the case of rabbits kept in hutches, and
plants grown under the shade or in a dark situation,
which have a less pronounced colour than when grown
in the sunlight. Gould was of opinion that birds of
the same species were more brightly coloured under
a clear atmosphere than when living near the coast
or on islands; Darwin informs us that there are few
British birds that are not dull coloured, while shells,
according to E. Forbes, are more brilliantly coloured
under a clear sky and in shallow water, than under a
dull atmosphere and at a greater depth. But whether

our view of the matter be correct or not, it is obvious
that the alternative colours of animals, or of plants,
cannot have been produced by natural selection. We
might just as well suppose that the moulting of the
feathers, or the shedding of the hair in old age, or the
sudden whitening of the hair of young persons, which
sometimes takes place in a single night, are due to the
same process.

According to Darwin, the first protective colours were
acquired by a happy physiological variation, but if, in
the first instance, the change had a physiological origin,
why not in all instances ? If Nature in one instance be
capable of producing a single individual with protective
colouring, she is surely able to do the same thing for
other individuals.   Why should we imagine that she has
put forth her last effort in producing one or more
individual specimens and no more, leaving it to natural
selection to continue her work ?   If we are not to suppose
this, then we must assume that Nature will go on producing other adapted variations in the same way, and
that new varieties and new species will continue to arise
in strict conformity with physiological laws.   Indeed,
this assumption is, in my opinion, absolutely necessary to
the successful action of natural selection, for unless
there be a constant succession of similarly modified forms,
as would be the case under uniform conditions, there
would be no chance of a stable variety or species being
established at all, as any occasional variations that might

appear would inevitably be lost by intercrossing, and all the more so if there were a tendency to indefinite variation. On the other hand, if we grant a tendency to vary in a definite direction, under definite conditions, there would be a constant succession of similarly modified forms, and the basis would thus be laid for new species. It is difficult to see how new species could otherwise have become established.*

The formation of instincts is assumed to be another of the achievements of natural selection. The claim is put forth by Darwin himself, and every true Darwinist endorses it. But first, what is instinct? An action, when it is performed by an animal without experience, especially by a very young one, and when performed by many individuals in the same way, without their knowing for what purpose it is performed, is, according to Darwin, an instinct.† There are many actions—habits and tricks of manner, for instance—which Darwin and others call instincts, which are improperly so termed. Such actions may be, and often are, hereditary, but they are not on that account to be classified as instinctive. Darwin, however, seems to think that all actions which are hereditary should be so classified, and he accordingly discusses at great length various singular habits and states of mind, which he calls instincts from the simple fact that they are hereditary. Thus, he says:—" Habitual

* *See* Appendix A. † *Origin of Species*, p. 205.

actions and states of mind do become hereditary, and may then, as far as I can see, most properly be called instinctive."* But although all instincts are hereditary, it by no means follows that whatever is hereditary is instinctive. In speaking of instinct, I shall, therefore, exclude all such habits, tricks of manner, and states of mind as are merely hereditary, and shall include among instincts those hereditary actions and affections only which are indispensable to the existence of the individual and to the maintenance of the race ; such as alimentation, self-preservation, reproduction, and parental affection.

It is difficult to ascertain Darwin's views on the origin of instinct, as his utterances are so contradictory. Thus he tells us that " if it can be shown that instincts do vary ever so little, then I can see no difficulty in natural selection preserving and continually accumulating variations of instincts to any extent that was profitable. And thus, as I believe," he goes on to say, " all the most complex and wonderful instincts have originated."† So far it would appear that Darwin considered that natural selection has here acted in the ordinary way—namely, by preserving existing and profitable variations. But

---

* Extract from Darwin's MSS. in *The Mental Evolution of Animals*, p. 264.

† *Origin of Species*, p. 206. Compare the following extract from Darwin's MSS. in the work of Dr. Romanes on *The Mental Evolution of Animals*, p. 264:—" I believe that most instincts are the accumulated result, through natural selection, of slight and profitable modifications of *other instincts*."

variations from what ? From other instincts ! Darwin is here professing to explain the *origin* of instincts, and he makes the innocent mistake of accounting for one instinct by tracing it to another. It apparently never once occurred to him that it was necessary to trace the original instinct to its source. True, he refers to habit as a possible factor in the formation of instincts, but he expresses no definite opinion on the subject. " I believe," he says, " that the effects of habits are in many cases of subordinate importance to the effects of natural selection of what may be called spontaneous variations of instincts ; " but immediately adds that " it can be clearly shown that the most wonderful instincts with which we are acquainted, namely, those of the hive-bee and of many ants, could not possibly have been acquired by habit;"* and he also agrees with F. Cuvier in comparing instinct with habit, and says that the comparison " gives an accurate notion of the frame of mind under which an instinctive action is performed, but not necessarily its origin."† In another place he tells us that if habit becomes inherited the resemblance between it and

* *Origin of Species*, pp. 205-207. Dr. Wallace says that Darwin "gave very little weight" to the view of instincts being acquired habits.—*Darwinism*, p. 441.

† Judging from the following extract, quoted from his MSS. in *The Mental Evolution of Animals*, p. 377, Darwin seems at one time to have believed that habit was a real factor in the formation of instincts, as he anticipates an objection to this view, and replies to it as follows :—" An instinct performed only once during the

instinct " becomes so close as not to be distinguished,"*
and further on he speaks of "habit *or* instinct," as if
they were one and the same. †
It is thus exceedingly difficult to get at his views on
this question. We are told that instincts originated in
natural selection by the preservation of other instincts;
that they originated also in habit; that habit when
hereditary is undistinguishable from instinct; and lastly,
that such habits are instincts. Darwin devotes a considerable amount of space to the discussion of this
question, and has a great deal to say about instincts of
a secondary character, which, however, throws no light
upon the origin of the primary or fundamental instincts,
with which alone we are concerned.

Darwin's failure to explain the origin of instincts on
the principle of natural selection has not deterred
some of his followers from renewing the attempt. Dr.
Romanes considers that instincts originated in natural

life of an animal appears at first sight a great difficulty on our
theory; but if indispensable to the animal's existence, there is no
valid reason why it should not have been acquired through natural
selection, like corporal structures used only on one occasion, like
the hard tip of the chicken's beak, or like the temporary jaws of the
pupa of the caddis-fly." In this passage we have exhibited a
singular mental process : instinct originates in habit ; but a habit
cannot be acquired by only one performance of an act ; nevertheless if the instinct be indispensable to the existence of the
animal, there can be no good reason why it should not have so
been acquired.

    * *Origin of Species*, p. 206.　　　　　† *Ibid.*, p. 209.

selection and habit, but, like Darwin, he concerns himself only with the former. " The first mode of the origin of instincts, and with this only we are concerned," he says, " consists in natural selection, or the survival of the fittest (*sic*), continuously preserving actions which, although never intelligent, yet happen to have been of benefit to the animals which first chanced to perform them. Thus, for instance, take the instinct of incubation. It is quite impossible that any animal can ever have kept its eggs warm with the intelligent purpose of hatching out their contents, so we can only suppose that the incubating began by warm-blooded animals showing that kind of attention to their eggs which we find is frequently shown by cold-blooded animals. Thus crabs and spiders carry about their eggs for the purpose of protecting them; and if, as animals gradually became warm-blooded, some species, for this or for other purposes, adopted a similar habit, the imparting of heat would become incidental to the carrying about of the eggs. Consequently, as the imparting of heat promoted the process of hatching, those individuals which most constantly cuddled or brooded over their eggs would, other things being equal, have been most successful in rearing progeny; and so the incubating instinct would be developed without there ever having been any intelligence in the matter." * Why, may we ask in passing, does the

---

* Article *Instinct* in *Encyclopedia Britannica*, ed. 1888. That this is Dr. Romanes' mature opinion on the subject may be

writer so carefully exclude the element of intelligence from his process? It is "quite impossible," he tells us to begin with, that any animal could ever have kept its eggs warm with the intelligent purpose of hatching out their contents, and he ends by triumphantly declaring that the whole process can be gone through "without there ever having been any intelligence in the matter." Dr. Romanes evidently wishes it to be clearly understood that natural selection is a purely mechanical process from first to last.

Dr. Romanes takes his readers a long way back, but he does not take them back quite far enough to account for the origin of the incubating instinct. This instinct, he tells us, began with the cold-blooded animals and was continued by the warm-blooded animals, the latter "showing that kind of attention to their eggs that we find is frequently shown by cold-blooded animals." This, we say, is going back a long way, but not far enough, for already he assumes what he undertook to prove. To begin with, he assumes the existence of the parental instinct when he affirms that his hypothetical animals showed "attention to their eggs;" next he assumes intelligence, notwithstanding his disclaimer, for does he not say that the cold-blooded animals carry about their eggs "for the purpose" of protecting them? Again,

assumed from the fact that he reproduces the above extract word for word in his subsequent work on *The Mental Evolution of Animals* before the incubating process could have begun there must have been material for it to operate upon—that is to say, there must have been eggs; the sexual instinct must also have been formed before eggs could have been produced; so that, in order to show how one instinct originated, Dr. Romanes assumes the existence of a whole series of instincts, all of them co-operating towards a definite end, yet all purposeless and non-intelligent. This can hardly be said to be the proper way to treat of the origin of things. Dr. Romanes should have begun at the beginning, and explained the origin of the first of the whole series of instincts which he started with.

Let us take his own standpoint, however. He affirms that this particular instinct originated in natural selection; but, according to his own showing, habit had something to do with it. Cold-blooded animals carry about their eggs for the purpose of protecting them, and as cold-blooded animals "gradually became warm-blooded, some species, for this or for other purposes, adopted a similar habit." "Some species"—were there then different species of animals in existence at this early stage, and did only "some" of these species adopt the habit of cuddling their eggs? What became of the other species which did no cuddling? And during the period which it took for the cold-blooded animals to become warm-blooded animals, what became of the eggs, and the egg-layers? How could the "species" continue to exist if the eggs were not hatched? But supposing that "some species" had chanced to discover that eggs required heat, they had still to learn how much was wanted, the number of hours, days, or weeks they would require to sit (for the eggs of some species take much longer time to hatch than others); they had also to discover that eggs required moisture and how much; that they had to turn them over regularly, so that each egg would receive its due amount of heat and moisture on all sides; and, before and above all this, these egg-laying and would-be egg-incubating animals would have to know (and how know except instinctively?) that when they had passed successfully through these various stages the eggs, if so manipulated, would develop into animals like themselves. And be it understood, all this experience would have to be acquired by each individual for himself, or rather herself, and not by the exercise of intelligence, which we have seen is carefully excluded, and acquired, be it also remembered, not gradually through long geological periods, as the writer assumes, or even during the lifetime of the individual, but in the course of the few hours during which the eggs will keep sound and fit for hatching. What possible chance would the poor egg-laying animals have of leaving any progeny behind them ?

Instincts are usually divided into two classes, namely, primary and secondary. The primary are those which are supposed to be indispensable to life. They are original, primordial endowments, and are contemporaneous with life itself. Secondary instincts, on the other hand, are acquired, and their origin in most, if not in all cases, can be traced to intelligence or habit. The use of artificial combs by hive-bees, the pointing and retrieving of dogs, the homing of pigeons, nidification, the fear of man, and the feigning of death, are instances of secondary or acquired instincts. The difference between primary and secondary instincts may be illustrated by the case of the Megapodes of New Guinea, which collect into a mound large quantities of leaves, grass, and other vegetable matter, in which they deposit their eggs, leaving them to be hatched by the heat generated in the mass by fermentation. This is a modified or acquired instinct; but had the bird not possessed the primary incubating instinct in the first instance it could never have acquired this modified form of it.

At the most, natural selection, as we have seen, only accounts for the origin of secondary instincts, and fails altogether in the matter of the primary instincts. There are two other theories of the origin of instinct which claim consideration; the first is that of Mr. Herbert Spencer, who advocates the theory of reflex action, and that of Lewes, who adopts the principle of lapsed intelligence. According to Mr. Herbert Spencer the order of development is—first, reflex action; secondly, compound reflex action; thirdly, instinct; and, last of all, intelligence. Lewes, on the other hand, begins with the last, and evolves instinct out of intelligence. Reflex

action is known by the uniformity of the reaction which
occurs on the application of any stimulus. In reflex
action the same reaction will take place whatever the
stimulus applied. The criterion of intelligence, on the
other hand, is that the reaction is adaptive or discriminative. Instinct, therefore, is not, according to this test,
reflex action, or any form of reflex action, for in instinct
there is invariably adaptation of means to ends, and Mr.
Herbert Spencer has not shown how any compounding of
reflex actions can make them discriminative. According to
Lewes, instinctive actions were originally intelligent, and
have become automatic from repetition, and ultimately
hereditary. There are many facts which give support
to this theory. Actions which at first are performed
with great difficulty become habitual and automatic
when often repeated. Thus, in learning to play the piano
a vast amount of labour has to be expended in learning
the notes, and in adjusting the fingers of the player to
the instrument, but an expert pianist will play a piece
without any effort and almost without consciousness.
And actions which become automatic in the individual,
also tend to become automatic in the race.

What we have to imagine is a condition of things in
which instant action follows sensation. This may be said
to occur in all cases of true or primary instinct—that is to
say, of instincts which are indispensable to life. The
peculiarity of instinct, and which is in so striking a con-

trast to reason, is that it is prompt in its action. It never deliberates, never wavers, never hesitates. The newly hatched chicken picks up the grain the instant it sees it; the newly-born foal makes straight for its dam's teats as soon as it can stand upright; ducklings hatched under a hen rush headlong to the pond in spite of the most frantic effort of their foster mother to prevent them; and the broody hen is perfectly incorrigible, and will persist in sitting, even if deprived of her eggs.

In man and in the higher vertebrates there is not only organic but also mental division of labour. The mental functions—sensation, perception, and volition—lead up to action; but experience shows that perception does not always follow sensation, nor volition perception, nor action volition. As often as not there is a break in the connections, or action may follow after a longer or shorter interval. But in instinct there never seems to be any break at all, or any interval between sensation and action; the perceptive faculty is in abeyance, and the organism appears to be under the control of an inner force which it is incapable of resisting. When instinct is strong, reason is weak. Dr. Wallace has remarked that the most perfect and the most striking examples of instinct are those in which reason and experience seem to have the least influence.* The invertebrata, in which mental activity is at the minimum, are almost entirely under the control of instinct; the lower classes of verte-

* *Natural Selection*, p. 201.

brata, which display more intelligence than the former, have less perfect instincts; among mammals instinct gradually diminishes till in man it reaches the minimum, while mental activity is at its maximum. In the cell or in unicellular organisms, which have no special organs and no division of functions, sensation, perception, volition and action would appear to occur simultaneously, forming, as it were, one instinctive act. Among the higher grades of organisms mental development corresponds with organic differentiation, while among the higher vertebrates, as in man, the perceptive faculties predominate over instinct. In the order of development, it would appear that instinct comes first, and reason afterwards; but reason never entirely supersedes instinct, even in man. Man is the creature of his instincts even in the exercise of his highest functions. His intuitions are not his in the sense of having been reached by him by any process of reasoning. They come to him he knows not how or whence. How could the orator move the masses if these had not an untaught logic to which he could appeal? There is a striking similarity between instincts and the higher manifestations of mind; our purest actions are impulsive; our best thoughts are intuitive. The fact that instinct is strongest when intelligence is weakest would not lead us to conclude that the former was developed out of the latter. Only on the supposition that the lower order of organisms originally possessed much higher intelligence than they appear to have now, can we imagine instinct to be lapsed intelligence.

Summary.—I have attempted to show that natural selection—understanding by that term the struggle for existence, or severe competition—can have no effect in producing new organic forms. For this purpose I have taken for illustration the two cases usually put forward by Darwinists in support of their views—namely, protective coloration and instinct, and I have endeavoured to show that neither of these can be regarded as the product of natural selection. So far as the latter is concerned, I have contended that Darwin and his followers have failed to prove that either the primary or the secondary instincts have originated in natural selection; while, as regards protective coloration, I have pointed out that it is not possible for an animal to acquire a seasonal or alternative colouring by that process, and that as protective coloration was, in the first instance, acquired through a physiological process, and without the aid of natural selection, there is no necessity for resorting to a mechanical process, such as natural selection, for its subsequent extension.

CHAPTER III.

THE EXTERMINATION OF THE UNFIT.

NATURAL selection is supposed to select the fit, but, at best, it only selects the unfit. Nature brings into the world more beings than she has made provision for, and, as all cannot live, she selects the unfit for destruction, and, according to Darwin, she uses the fit as her instrument for that purpose. "As in each fully stocked-up country natural selection acts by the selected form having some advantage in the struggle for life over the other forms, there will be a constant tendency in the improved descendants of any one species to supplant and exterminate in each stage of descent their predecessors and their original progenitors."* And this struggle for life which spares neither predecessors nor progenitors is said to be most severe between allied species or near relations, "as having the same structure, constitution, and habits,"† and presumably consuming the same kind of food, and, as a necessary consequence of this, we are told that "all the intermediate forms between the less and the more improved states of the same species, as well as the parent species itself, will generally tend to become extinct."‡ Do we really find that the "improved

* *Origin of Species*, p. 93.   † *Ibid.*, p. 86.   ‡ *Ibid.*, p. 93.

descendants" exterminate their predecessors, and that all
"intermediate forms" of the same species become extinct?
Darwin's hypothesis requires that such a result should
take place, but he has supplied us with no proofs in
support of it. The fact is, all the evidence proves the
contrary, for not only do intermediate or less improved
forms exist now side by side, but they appear to have
so existed from the remotest ages.

Thus, if two kinds of organisms are competitors for the
same means of subsistence, the more highly organized
being would, according to Darwin, prove itself the more
fit of the two, and would exterminate the less highly
organized being. If, therefore, we can show that two
species, or two varieties of the same species, the one
more highly organized than the other, and both com-
peting for the same subsistence, are yet living side by
side, then we shall have a clear case of the survival of
the unfit under natural selection.

The *Foraminifera* furnish just such an instance. In
some of the groups of this order the shell consists of a
single chamber, and the animal is nothing more than a
little mass of protoplasm enveloped in a calcareous covering. In the more complex groups the protoplasm under-
goes a subdivision into partially separated segments. In
these polythalmous varieties the shell consists of a series
of chambers, separated by the partitions of the test, and
filled with sarcode. The forms assumed by this order
are extremely diversified, and in regard to size they vary

from a mere microscopic speck to an inch or more in diameter. Now, all these various forms are found to exist together at the present day in all oceans, in almost all seas, in shallow water as well as at great depths, and these various forms lie so contiguous that a single dredging will often bring up several distinct species and genera. They appear also to have been quite as numerous in former times, as they abound alike in Paleozoic, Mesozoic, and Kainozoic formations. Whole beds of the carboniferous limestone in Europe, Asia, and America are composed of the shells of the *Foraminifera*; and in the secondary rocks they also occur in great abundance, the formation known as the chalk being almost wholly composed of them. These organisms are therefore not only amongst the oldest, but they are, and have been during vast geological periods, the most widely distributed in the world. Now, it cannot be supposed that the larger and more complex structures had any advantage in the struggle for existence over the smaller and simpler structures, for, as has been pointed out,* they are as fully incapable of escaping from their enemies by any movement as the latter, and show no difference in aspect such as would enable them to escape observation, but rather the contrary, as their greater size renders them more conspicuous to their enemies. May we ask then, where, in the case of the *Foraminifera*, is the extermination of the unfit ? In other words, does not

* *Nature and Man*, by Dr. Carpenter, p. 458.

the continued existence of low forms of animal life in the
same locality, and presumably subsisting on the same
kind of food, show that natural selection has failed in
its operation?

Darwin anticipates our argument, and evades it in a
characteristic fashion. "What advantage," he asks,
"would it be to an infusorian animal—to an intestinal
worm—or even an earthworm, to be highly organized?
If it were no advantage, these forms would be left by
natural selection unimproved, or but little improved,
and might remain for indefinite ages in their present
lowly condition."* Why does he not finish his sentence
by stating the alternative? Suppose it *were* an advantage, what then? If some of these lowly organized
animals were improved and others not, are we to be told
that the improved individuals would not have any advantage over the unimproved? Have we not been
informed that natural selection leads to "the extermination of the less improved and intermediate forms of
life,"† and that "the very process of natural selection
implies the continual supplanting and extermination of
preceding and intermediate gradations?"‡ Dr. Wallace
has also a reply. He tells us that low forms exist because they "occupy places in nature which cannot be
filled by higher forms," and because they have "few or
no competitors."§ But this does not meet the case we

* *Origin of Species*, p. 98.   † *Ibid.*, p. 103.
‡ *Ibid.*, p. 164.   § *Darwinism*, p. 114.

have cited, for the lower forms of life not only exist, but
they exist side by side with higher forms, and that they
are not without competitors must be obvious, as every
gradation of form is represented, and, according to
Darwin, it is just among individuals of the same or of an
allied species that the struggle for existence is most
severe and destructive.*

A superior race does not usually exterminate an
inferior one. The former seldom pushes his victory to
the point of extermination, while the latter, after a trial
of strength, accepts his defeat with good grace, and sub-
mits to take a subordinate position, the contest generally
ending in the amalgamation of the two races. This, at
all events, is what happens with the different races of
mankind, and with plants and animals of the same or of
an allied species. Everywhere we find them living
together in the same place, and subsisting on the same
kind of food. Even Carnivora do not fly at each other's
throats whenever they chance to meet, although, knowing the consequences of an encounter, they are careful to
keep at a respectful distance from each other. But if
Darwin's contention be correct, we would never find an
inferior and superior race living together; in every
locality there would only be one species, and that the
most highly organized ; and thus a few superior races
would partition the earth amongst them to the entire

* *Origin of Species*, pp. 59, 86.

exclusion of the innumerable varieties, species, genera, and orders which now inhabit it.

On this question of the extermination of races Malthus is a safer guide than Darwin. Epidemics, improvidence, and disease—"famine, pestilence, and sudden death"—are what have struck terror into the heart of mankind in all ages, because they have been the most destructive to human life, and it is to such catastrophes as these that Malthus ascribes the periodical decimination in times past of the population in certain countries. "As savages," he says, "are wonderfully improvident, and their means of subsistence always precarious, they often pass from the extreme of want to exuberant plenty, according to the vicissitudes of fortune in the chase, or to the variety in the produce of the seasons. Their inconsiderate gluttony in the one case, and their severe abstinence in the other, are equally prejudicial to the human constitution, and their vigour is accordingly at some seasons impaired by want, and at others by a superfluity of gross aliment and the disorders arising from indigestion. These, which may be considered as the unavoidable consequence of their mode of living, cut off considerable numbers in the prime of life. They are likewise extremely subject to consumptions, to pleuritic, asthmatic, and paralytic disorders, brought on by the immoderate hardships and fatigues which they endure in hunting and war, and by the inclemency of the seasons, to which they are continually exposed. The missionaries speak of the Indians of South America as subject to perpetual diseases, for which they know no remedy."*

There is no case on record where one race has deprived another of the means of subsistence. In recent years the superior and inferior races of mankind have come in contact all over the world, to the manifest advantage of the latter, so far at least as the means of existence are concerned. For proof of this, one has only to read the narratives of Cook, Vancouver, and La Pérouse, and compare them with the accounts of modern travellers in the same regions. It is true that many of the aboriginal races are dying out; but this result is not due to any antagonism of the superior races towards the inferior, or to any competition for the means of subsistence. Their disappearance is entirely the result of constitutional causes. Almost everywhere the aboriginal races exhibit a strong tendency to contract disease, and an inability to throw it off when once contracted. It has been remarked that the natives of the South Pacific Islands on the Queensland sugar plantations give up all hope of recovery as soon as they are attacked by any illness, and never make any attempt to throw it off, and as a natural consequence many of them die from slight attacks which Europeans would get rid off without any trouble. Contact with civilized man seems also to have a disastrous effect on savage races in various ways. Bates, for instance, men-

* *Principles of Population*, p. 62, 5th ed.

tions a tribe of Indians on the Amazon—the Pasees, which he describes as a gentle, industrious, and intelligent people, which were formerly most numerous, but are now almost extinct. He says :—

The principal cause of their decay in numbers seems to be a disease which appears amongst them where a village is visited by people from the civilized settlements—a slow fever, accompanied by the symptoms of a common cold (" defluxo," as the Brazilians call it), ending probably in consumption. The disorder has been known to break out when the visitors were entirely free from it ; the simple contact of civilized man, in some mysterious way, being sufficient to create it.*

It is possible to conceive that the antagonism between Carnivora and Herbivora might lead to improvement of a functional kind, such as increased alertness, speed, and combativeness, the effect to some extent of the law of use and disuse ; but any struggle between the same or between allied species is almost invariably injurious to the individuals engaged in it. It has become the fashion amongst political economists of a certain school to speak in unqualified laudatory terms of the advantages of competition ; but no authority in political economy has yet hazarded the assertion that competition benefits the competitors. That it often is of advantage to the consumers is beyond a doubt ; but it is difficult to see the gain to the competitors. As a rule, competitors when severely pressed call for a truce ; they combine to reduce

* *The Naturalist on the Amazons*, p. 257.

5

the output and to raise prices, so that severe competition in the end does not invariably benefit even the consumers. The struggle for existence cannot possibly benefit competitors on both sides, while it is probable that both will suffer from it. A labour strike is a struggle between workmen and employers, but, whoever wins in the end, it is certain that both suffer as long as the struggle lasts, the workmen in wages and the employers in profits. Traders compete in order to secure a market, animals in order to obtain the means of subsistence; but unless the struggle comes to an end by the retirement or extermination of one of the competitors no benefit will accrue to either side. For the full development of any organism there is required the proper quantity of nourishment, and the due amount of exercise of all the organs and the abuse of none. In the struggle for existence all these conditions are absent, and in place of these there is privation and suffering in every case, even to the survivors, and if the struggle be protracted the result will be stunted growth, disease, premature decay, and a weakened constitution to transmit to posterity.

What is the general tendency of natural selection? Are its effects relative or absolute? Does it lead to the general advancement of the individuals subjected to its influence, or has it sometimes a contrary effect? Here, again, we find Darwin unsatisfactory. He says:—
" There is no good evidence of the existence in organic

beings of an innate tendency towards progressive development, yet," he adds in the same sentence, "this
necessarily follows . . . through the continued
action of natural selection."* If it "necessarily follows,"
surely that is good enough evidence. Again, we are
told that "the ultimate result" of natural selection is
"that each creature tends to become more and more
improved in relation to its conditions. This improvement inevitably leads to the gradual advancement of
the organization of the greater number of living beings
throughout the world."† But if the gradual advancement of organization is the "inevitable" result the improvement would be more than relative—it would be
absolute; and here again he is confronted with the
principle laid down by himself, and repeatedly insisted
on—namely, that structures are adapted to the conditions
of life and may therefore occasionally assume degraded
forms.

But if Darwin speaks hesitatingly on this subject, Mr.
Herbert Spencer's generalization is, as usual, sweeping
enough. Speaking of the struggle for existence, he
says:—" The pressure of population on the means of subsistence has produced the original diffusion of the race.
It compelled man to abandon predatory habits and take
to agriculture. It led to the clearing of the earth's
surface. It forced men into the social state; made social
organization inevitable, and has developed the social

* *Origin of Species*, p. 176.   † *Ibid.*, p. 97.

sentiments. It has stimulated the progressive improve-
ments in production, and to increased skill and intelligence. It is daily thrusting us into contact and more
mutually dependent relationships." * We cannot accept
such an optimistic view, even from Mr. Spencer. We
draw a distinction between incentives to action and the
struggle for existence, between the hope of reward and
fear of punishment. The former is stimulating, the
latter depressing in its effect. So far from being an
incentive to civilization, the struggle for existence has,
in my opinion, rather been a hindrance to it. What
possible chance has the savage of improving himself,
physically or mentally, if his whole time and all his
energies are engaged in a ceaseless endeavour to procure
the bare necessaries of life? Even if he succeeded in
occasionally satisfying his hunger, how could he take
even the first step towards an improved social condition
without entering into some sort of tribal contract, and
so, to that extent, abandoning the principle of competition? Except by concert with others, how could he
secure the benefits of division of labour, which is as
necessary to social as the separation of functions is to
organic progress? Such advantages are not to be obtained by competition, but by co-operation, which is a
principle of an altogether different kind. An agricul-
tural life, which is so essential to social advancement, is
impossible among rude tribes about equally matched

* *Principles of Biology*, part vi., p. 376.

(when competition is most severe), and constantly at war with each other. Before a man will sow he must see some probability that he will also reap. An increase in strength in one tribe gives it an advantage in contending with another tribe, while a decrease in strength at once lays it open to the attack of its enemies. It is only when the principle of co-operation is carried out on a large scale, as by the amalgamation of numerous tribes, and the community is thus enabled to protect itself, that opportunity is afforded for social advancement.

Were the struggle for existence as beneficial as it is represented to be we should expect to find a high state of civilization in those countries where the inhabitants were subjected to the severest form of competition. Do we find it so, as a matter of fact? Did the various tribes of North America, for instance, who, before the advent of the white man amongst them, were always at war with each other, ever attain a high state of civilization? The natives of Terra del Fuego, who cannot find time even to provide their bodies with covering against the rigorous climate of those regions, so incessantly are they occupied in attempting to procure even the barest supply of food, are they not the most wretched specimens of the human race in existence?

I have said that it would be impossible for civilization to make a beginning under pressure of severe competition; I go further, and maintain that an existing civilization under such conditions, so far from progressing, is more

likely to retrograde.   No one will dispute that the
Chinese race are inferior to the Anglo-Saxon.   Yet when
those two races come into competition, as they have in
California and in Australia, the result has been detri-
mental, not to the former but to the latter; and a
moment's consideration will show that it could not be
otherwise.   The Anglo-Saxon has more wants, or, what
is the same thing, has a higher standard of comfort, and
a higher social and intellectual position to maintain than
his Chinese rival, and he cannot therefore work at the
same rate of wages, or for the same number of hours as the
latter.   The Chinese labourer lodges in a hovel, spends
little on food, which is of the simplest and cheapest kind,
and less on clothing, and as his intellectual resources are
limited, he finds it little hardship to work for long hours at
low wages.   When two such races come into competition
one of two things must happen : either the Anglo-Saxon
must lower his standard of living, and relapse into a
lower social and intellectual position, or he must emigrate,
or starve.   If he accepts the first alternative he will
lower himself physically, morally, and intellectually, to
the condition of his rival ; if the latter, his rival will
supplant him, and thus prove himself to be the "fittest"
for that condition of existence.   One has only to
look at the wretched condition of the shirt-makers
and tailors of the East-end of London to obtain
evidence of the pernicious effects of severe competition.   We do not find that the struggle for

existence has had an elevating influence upon this class of people, but we have proof in abundance of the evil effects of adaptation to degrading and depressing surroundings.

Do we, then, assert that there has been no extinction? By no means. There has been extinction on a vast scale, as the geological records testify; but we have yet to ascertain the cause of this. According to Darwin the extinction of species is wholly, or almost wholly, due to the struggle for existence. I venture to dissent from this opinion. Here, however, it will be necessary to have a clear understanding as to the meaning we attach to the term "struggle for existence." The struggle for existence is of two kinds: organisms struggle with one another, but they also struggle against the adverse forces of nature. The former is competition; the latter is a struggle for existence, properly so called. It is the latter rather than the former that, in my opinion, is the chief cause of the extinction of species. Nor is it necessary that we should have to resort to the theory of great cataclysms or extraordinary convulsions of nature to explain the phenomena, as, I believe the ordinary climatic and geological changes that are going on at the present time are quite sufficient to account for the great proportion of the destruction of life that has occurred on the earth. Take, for instance, sudden variations of temperature. These have had a most disastrous effect on all kinds of organisms. The change from a very warm to a cold temperature, or, conversely, from an extremely cold to a warm temperature, is uniformly attended with great loss of life. Darwin mentions a case within his own experience, when in one severe winter (1854-5) he discovered that four-fifths of the birds on his grounds had been destroyed by the cold, and I am informed that the excessive heat of the present year (1890), in the southern portion of Australia, has had a very destructive effect on the parrot tribe. Indeed, so impressed was Darwin by the injurious effect of climatic changes that he distinctly states that " periodical seasons of extreme cold seem to be the most effectual of all checks in determining the numbers of a species."*

But does not competition also lead to the extinction of species ? No doubt it does ; but, in my opinion, only to a limited extent. If one species be better endowed than another species, it will undoubtedly have an advantage in the struggle for life ; but it will seldom happen that the less gifted species has not a set-off of some kind, such as greater prolificacy, or greater numerical strength, which will enable it to hold its own against a race the individuals of which are stronger or better armed. And if the competitors belong to the same or to an allied species, the improved individuals, being numerically inferior, will be absorbed by the less improved but numerically superior section of the species. On the other hand, if a species be suddenly

* *Origin of Species*, p. 54.

confronted by a new and better gifted competitor or
enemy, whether of the same or of a different species
or genus, the consequences are likely to prove serious
to the less gifted species—witness the progress of the
Norwegian or brown rat, which has almost exterminated
every variety of rat in Europe. The advent of a new
competitor, or a new enemy, usually proves disastrous
when it appears suddenly on the scene, and before its
victims have had time to adapt themselves to their new
relations.

Summary. — I have here attempted to show that
natural selection, or the struggle for existence, is not competent to preserve variations, the struggle for existence
not being in its nature preservative; that an organism
does not survive in consequence of the struggle, but
because it is adapted to the conditions of life, as otherwise
the greater the struggle the greater would be the chance
of survival, whereas the unfit probably struggle more
than the fit. Natural selection is a purely destructive
process. Nature brings more beings into existence than
she has made provision for, and she preserves the fit
only by destroying the unfit. Natural selection is of
two kinds : in the one case there is a struggle against
the adverse forces of Nature, in the other there is a
struggle between the organisms themselves. This latter
is competition. Darwin informs us that competition is
always more severe between varieties of the same or of

allied species—that is, between near relations—because they are competitors for the same kind of food, and consequently there is greater destruction amongst the latter than amongst species less closely related. He also maintains that the more improved form of a species exterminates the less improved; but he has produced no evidence in favour of this view, and we have shown, as in the case of the *foraminifera* and the human race, that this theory cannot be maintained, as in these instances we find the intermediate, or less improved, forms of the same species existing side by side with the more improved. It would rather appear that a less improved variety, being the more numerous at the start, absorbs the more improved, to the benefit of the former. The chief cause · of the extinction of species seems to be sudden changes of climate, and, in a lesser degree, the sudden appearance of an enemy not indigenous to the country, and for which the natives were unprepared. If climatic and other changes were brought about gradually, and if the species had only to contend against enemies which were indigenous to the country, extinction would proceed on a much less extensive scale than it appears to have done in the past, as the organisms would then have time to adapt themselves to their new conditions and their new enemies.

CHAPTER IV.

SEXUAL selection, according to Darwin, is not a struggle between the sexes, but between individuals of the same sex. " The sexual struggle," he says, " is of two kinds ; the one is between individuals of the same sex, generally the males, in order to drive away or kill their rivals, the females remaining passive, whilst in the other the struggle is between the individuals of the same sex, in order to excite or charm those of the opposite sex, generally the females, which no longer remain passive, but select the more agreeable partners.* The superior strength and the weapons of offence possessed by the male have therefore, according to this view, no direct relation to the opposite sex, whereas the personal attractions of the male, such as bright colours, plumage, and so forth, have direct relation to the opposite sex, having been acquired for the purpose of alluring the female, and the selection of a partner, after all the efforts of the male to allure the female is, at the last, left to the weaker sex.

Darwin also contends that secondary sexual characters are transmitted by the male to both sexes, the male,

* *Descent of Man*, 2nd ed., p. 209.

being, in his opinion, more variable than the female
(which is surely contrary to all human experience), is
modified first and subsequently transmits his modified
form to both sexes.* Secondary sexual characters, he
informs us, comprise " the greater size, strength and pugnacity of the male, his weapons of offence or means of
defence against rivals, his gaudy colouring and various
ornaments, his power of song, and other such characters."†
Many facts and arguments are adduced by Darwin,
some of them of a very singular kind, in support of his
theory of the transfer of secondary male characters to
the female. He tells us, for instance, that " when the
male is brilliantly coloured, and differs conspicuously
from the female, as with some dragon-flies and butterflies, it is probable he owes his colours to sexual selection, while the female has retained a primordial or
very ancient type of colouring. . . . In other cases
in which the sexes resemble each other and are both
brilliant, and especially when the colours are arranged
for display, we may conclude that they have been
acquired by the male sex as an attraction, and have
been transferred to the female."‡ If we admit the probability of the latter statement we should not be asked
at the same time to admit the former ; and if we assume

* *Descent of Man*, 2nd ed., p. 319.
† *Ibid.*, pp. 208, 471, 543, 631-4, and elsewhere ; in fact, the
greater portion of the *Descent of Man* is devoted to the proving of
this point.
‡ *Ibid.*, p. 328.

that when the sexes are both brilliantly coloured, the brilliant colouring "having been acquired by the male sex as an attraction," we shall have to ask what becomes of the attraction when it has been "transferred to the female?" Again, he says:—"In no case have we sufficient evidence that colours have been thus acquired [by sexual selection] except when one sex is much more brilliantly and conspicuously coloured than the other, and when there is no great difference in habits between the sexes sufficient to account for their different colours. But the evidence is rendered as complete as can be only when the more ornamented individuals, almost always the males, voluntarily display their attractions before the other sex; for we cannot believe that such display is useless, and if it be advantageous, sexual selection will almost invariably follow."* We should have imagined, on the contrary, that when one sex, say the male, is much more brilliantly or conspicuously coloured than the female, it was a proof that this colouring was advantageous to him in a sexual point of view, and, being so, should not be transferred to the opposite sex, otherwise he would cease to be attractive, and would, therefore, lose any advantage which he might have possessed. The fact that the most ornamented males display their attractions before the females shows that they consider them advantageous in their sexual relations; and if they are advantageous in this respect

* *Descent of Man*, p. 260-1.

that is a sufficient reason why they should not be transferred to the other sex. The very fact of the male displaying his attractions before the female, in our opinion, renders the evidence "as complete as can be" that no such transfer has taken place, or otherwise the purpose for which they had been acquired would be defeated. According to the principles of sexual selection (and, of course, the same holds good of natural selection) no character could be preserved by one sex unless it were in some way advantageous to that sex, and it is evident if it were transferred to the opposite sex it could no longer be an advantage to the sex which originally acquired it. If we imagine the muscular form, the sonorous voice, and hirsute appendages of the man transferred to the woman, we would better understand the nature of the transformation which Darwin alleges has taken place in the animal world generally. Secondary sexual characters, which are becoming in the man, would be hideous in the woman.

The intimate connection between the reproductive organs and secondary sexual characters, as proved by the effect of castration, also precludes the possibility of any transference. All the secondary sexual characters of the male disappear, or are greatly modified, after this operation has taken place. If performed on the male early in life, secondary sexual characters never appear at all. On the other hand, all these characters are intensified during the breeding season—colours become brighter, plumage more brilliant, odours more pungent, and the voice sweeter or louder—while some animals, like the sea-elephant and the bladder-nose seal, acquire extraordinary vocal appendages, which appear only in the males, and in the males only during that season.

The question we have to determine is, for what purpose have secondary sexual characters been acquired by the males? For the purpose, says Darwin, of driving off or killing their rivals, and for alluring the females. But this explanation will not apply in all cases, as we shall see when we ascertain what these secondary sexual characters are.

There is, first, size and strength. The male is almost invariably both larger and stronger than the female. In many species the male is very much larger than the female, especially among polygamous animals—as, for example, among whales and seals; in one of the latter species, *Callorchinus ursinus*, the male is as much as six times larger than the female. The exceptions to this rule are far from being numerous, and comprise certain species of insects, fishes and birds.* It is a remarkable

* Among birds Darwin includes the emu, the females being, he says, larger and more pugnacious than the males, the latter also performing all the duties of incubation, and have even to defend the young from the fury of the mother, who is represented as a perfectly ferocious character. In confirmation of this, he quotes the following observations of Dr. Bennett, of Sydney, on the habits of this bird:—" As soon as she catches sight of her progeny she becomes violently agitated, and notwithstanding the resistance of

fact that when the male is smaller or weaker than the
female he is also effeminate in his character, while the
larger or stronger female has all the pugnacity of the
male.

Weapons of offence or defence are also, as a rule, peculiar
to the male, such as horns, tusks, incisor teeth, spurs,
and so forth. Among many families of beetles the males
are furnished with extraordinary horns projecting from

the father, appears to use her utmost endeavours to destroy them.
For months, also, it is unsafe to put the parents together, violent
quarrels being the inevitable result, in which the female generally
comes off the conqueror." It is almost a pity to spoil such an
interesting story, but truth compels me to say that it is not
founded on fact. I learn from Mr. Richard Bennett, of Victoria,
a gentleman who has spent nearly half a century in the Australian
bush, and has carefully watched the emu in its native haunts, that
"the male differs in no respect from the female, and that the
latter is very attentive to her young." Mr. Le Souëf, Director of
the Melbourne Zoological Gardens, has also kindly supplied me
with the following information :—" We have a good many of these
birds here. One pair bred regularly every year, last season rearing
five young ones. My experience of these birds agrees with Mr.
R. Bennett's. The male and female birds sit alternately, and
when the young are hatched, the female takes the most care of
them. The male is the more pugnacious of the two, and is unsafe,
especially with children, during the breeding season. The male
has a little more colour about the head than the female, and has
also a long bunch of feathers on the breast, which it is fond of
displaying ; but otherwise it is most difficult to tell a male from a
female when apart, the male being slightly the larger of the two.
A few years ago we had a female bird killed by a leopard while
sitting on her eggs, and the male bird took her place and hatched
the young out unassisted and reared them up."

the forehead. in some cases longer than their bodies—as. for instance, the *Lamellicorns*—which are entirely absent in the females. The teeth of the male salmon are sharp-pointed, while those of the female are broad and flat. The males among reptiles have usually very formidable weapons, as in the case of *Cameleon Owenii*, which has a snout and three horns as well, of which the female has not a trace. The males of most species of birds are provided with powerful beaks and spurs, the latter being absent in the females. Among deer the stags only have horns, the females being entirely destitute, except in the case of the reindeer. Among many of our domestic breeds of goats and sheep, the males alone have offensive weapons, while among almost all wild animals of the same species the females have horns, but they are much smaller than those of the males. Stallions have also canine teeth, which are either smaller or altogether absent in the mares. These teeth are freely used on the mares during the breeding season. Many male ruminants also possess canine teeth, of which the females are destitute, which are probably used for the same purpose.

Colour is another sexual character. The males are almost always more distinctly marked than the females; sometimes, as in the case of birds, they are gorgeously coloured, while the females are plain. Among *Diptera* the males are darker than the females, as are also spiders, moths, and field-bugs (*Hemiptera*). Male beetles are often splendidly coloured, as are also male butterflies

6

especially in the tropics, while the females are plainly marked. Among fishes the male is usually more brilliantly coloured than the female, especially during the breeding season. Everyone knows how conspicuous are the colours and plumage of the male pheasant, and the marked contrast between the sexes in this respect. The males of the Antarctic goose, the silver pheasant, and the bell-bird of South America are white, while the females are obscurely coloured. Among mammals the colour of the male is usually more pronounced than that of the female.

The sexes also differ in the sounds they utter. These are of two kinds—pleasant and harmonious, and harsh and discordant. Among *Homoptera* and *Orthoptera* the males alone possess the power of song, or make stridulating sounds. Among birds the vocal organs are developed to their greatest pitch, the male being usually the best songster. Almost all mammals make frequent use of their voice during the breeding season, especially the males, and the male voice is usually much louder and harsher than that of the female.

Odour is another male sexual character almost peculiar to the male. In many animals the males alone possess scent glands, and in those cases where the sexes have them in common they are more highly developed in the males.

Generally speaking, when the sexes do not differ in any of the foregoing characters, the male possesses some other peculiarity, either by way of ornament, distinction, or utility, which is absent in the female. But in addition to these the males of some species of animals have secondary sexual characters of a very special order. The males of certain species of ants, for instance, possess wings, which Darwin supposes to be used for the purpose of enabling them to pursue the females. These appendages are, of course, peculiar to the male. The males of some species of beetles have instruments of various kinds for seizing and holding the female when she is reluctant. These, also, are peculiar to the males, and under no circumstances can we imagine them to be transmitted to the other sex. Although Darwin in numerous places refers to these organs and explains their uses, he seems to have overlooked their bearing on the question before us. Speaking of insects, he says:—"The mandibles or jaws are sometimes used for this purpose [that is seizing the females]; thus the male *Corydalis cornutus* has immense curved jaws, many times longer than those of the female, and they are smooth instead of being toothed, so that he is thus enabled to seize her without injury. One of the stag beetles of North America (*Lucanus elaphus*) uses his jaws, which are much larger than those of the female, for the same purpose, but probably likewise for fighting."* Again, referring to the extraordinary size of the horns of other beetles, he observes that "their widely different structure in closely allied

* *Descent of Man*, p. 275.

forms indicates that they have been formed for some
purpose; but their excessive variability in the males of
the same species leads to the inference that this purpose
cannot be a definite one. The horns do not show breaks
of friction, as if used for any ordinary work. Some
authors have supposed that as the males wander about
much more than the females, they require horns as a
defence against their enemies; but as the horns are often
blunt, they do not seem well adapted for defence. The
most obvious conjecture is that they are used by the
males for fighting together, *but the males have never
been known to fight,* nor could Mr. Bates, after careful
examination of several species, find any evidence in their
mutilated or broken condition of their having been thus
used."* Darwin brings a long array of facts to prove
that the females, especially among birds, show a preference for certain males, but the authorities he quotes are
not with him in this view. Mr. Brent, for instance, gives
it as his opinion that " old hens, and hens of a pugnacious
disposition, dislike strange males, and will not yield
until beaten into compliance."

Darwin is probably correct when he states that certain sexual characters have been acquired by the male
for the purpose of charming the female, amongst which
may be included bright colours, handsome plumage,
wattles, tufts, and so forth, and also the power of song,
and probably pungent odours; but why should we

* *Descent of Man,* p. 297.

suppose that size, strength, and weapons of offence should have been acquired by the males for the purpose of driving off or killing their rivals ? All the other male sexual characters have relation to the opposite sex, and why not these also ? The secondary sexual characters of the male may be divided into two classes : those which are attractive, and those which are dominant. To the former belong brilliant colours, ornaments, odour and song; to the latter size, strength, and weapons of offence. Vocal sounds belong to both divisions : those which are soothing and musical being attractive, and those which are harsh and dissonant being dominant. Male sexual characters are, in their very nature, co-related with the opposite sex, otherwise they cannot be regarded as sexual characters at all. And why, for instance, should the males be endowed with greater physical strength and with weapons of offence for the purpose of fighting with other males ? It is evident that if all the males of the same species were equally well armed, no advantage would be gained, as all would be equally matched ; the only difference would be that if they were powerful and well armed they would be able to inflict greater injury upon one another. There would, however, be good reason for the males possessing these characters if we supposed them to have reference to the opposite sex. The female is known to be coy, variable in temper, and often reluctant to accept the advances of the male. As these peculiarities of the female might prevent the propagation of the species, and the propagation of the species being the main object for which the sexes exist, it seems not altogether improbable that the dominant characters of the male have been acquired for the purpose of overcoming the reluctance of the female.

Darwin's theory assumes that an advantage would be gained if the stronger and better armed males killed off the weaker and worse armed, and that it is the object of sexual selection to bring about this result. In such a case we should expect to find everywhere fewer males than females, since the latter, not having been subjected to this species of natural selection, would not be diminished in numbers, while the former would; but Darwin admits that he can find no satisfactory evidence that such is the case.* But this is a very one-sided sort of an argument altogether, for, if the weeding out of the inferior males is an advantage to the race, I fail to see why the same process should not be extended to the females. If size, strength, and beauty be the results of the struggle between the rival males, why should not the females be allowed to participate in it also, and thus reap the same advantages ?

Nothing could better illustrate the purposes for which the males of certain animals use their strength than the following extract from Captain Bryant's narrative, quoted by Darwin. Speaking of the polygamous eared seals, he says :—

* *Descent of Man*, p. 282.

Many of the females, on their arrival at the island where they breed, appear desirous of returning to some particular male, and frequently climb the outlying rocks to overlook the rookeries, calling out and listening as if for a familiar voice. Then, changing to another place, they do the same again. . . . As soon as the female reaches the shore, the nearest male goes down to meet her, making meanwhile a noise like the clucking hen to her chickens. He bows to her and coaxes her until he gets between her and the water so that she cannot escape him. Then his manner changes, and with a harsh growl he drives her to a place in his harem. This continues until the lower row of harems is nearly full. Then the males higher up select the time when their more fortunate neighbours are off their guard to steal their wives. This they do by taking them in their mouths and lifting them over the heads of the other females, and, carefully placing them in their own harem, carrying them as cats do their kittens. Those still higher up pursue the same method until the whole space is occupied. Frequently a struggle ensues between two males for the possession of the same female, and both seizing her at once pull her in two, or terribly lacerate her with their teeth. When the space is all filled, the old male walks around complacently reviewing his family, scolding those who crowd or disturb the others. This surveillance always keeps him actively occupied.*

This is undoubtedly a rough sort of courtship. The attractive and the dominant sexual characters of the male are used by turns, the tender clucking suddenly changing into a harsh growl according to circumstances. Having filled his harem in this rude fashion the male seems to have had ample opportunities afforded him afterwards of exercising his marital power in establishing order among his numerous wives.

* *Descent of Man*, p. 523.

Curiously enough, primitive marriage amongst mankind was not much different from this. Either the
female was forcibly abducted by the husband from
another tribe, or she was at the absolute disposal of her
nearest male relative, who gave her in marriage to whom
he chose, without consulting her wishes in the smallest
degree. If she refused to accept the husband offered to
her, she was simply beaten into compliance. The treat-
ment the females have received from time immemorial at
the hands of the stronger males seems to have had a permanent effect on the mental character of the weaker sex,
as they will continue to be attached to men who use them
ill, but whose brutality goes along with power, more than
they will continue to be attached to weaker men who use
them well.*

If the purposes for which secondary sexual characters
have been acquired by the male be what we have described,
there can be little difficulty in deciding as to the likeli-
hood or otherwise of their being transferred to the
opposite sex. I do not think the transfer probable, or
even possible, without a reversal of the order of nature.
It is true there are instances where the females of some
species of animals have acquired secondary sexual
characters, as in the case of the domestic fowl, for instance ; but this only occurs when the female has become
old or diseased, and such characters are not transmitted.

* Spencer's *Study of Sociology*, p. 377.

One of the most singular of Darwin's conclusions is, that it is the female that selects the male, and not the male that selects the female. The male, he alleges, is too ardent to make a selection, so he leaves it to the more passive female to choose her mate. One would imagine that the very eagerness of the male shows that he is the active agent in selection. Why otherwise does he display his attractions before the females ? Why does he drive away or kill his male rivals ? And if the stronger or better armed males succeed in driving off or killing the weaker or worse armed, what choice is there left for the females ? But, unless mankind are to be excluded from the operation of sexual selection, what is the lesson which human experience teaches in this respect ? Is it the male or the female that selects in this case ? Nor are the males among animals without their likes and dislikes, as anyone may see who carefully observes the habits of our domestic animals. Stallions, for instance, will often reject certain mares which are brought to them, and take to others, as every breeder knows. Turn a stallion into a paddock with a number of brood mares during the season, and he will cut out a few whom he favours, and will drive off the rest, and will not even allow the latter to come near his favourites.

Darwin regards sexual selection as supplementary to natural selection, but the two theories do not harmonize. Natural selection looks only to the present, never to the future ; to immediate utility, never to future require-

ments ; to the relative, never to the absolute. Variations
are appropriated when advantageous for the time being,
but a variation that enables an animal to escape the
observation of its enemies, or to subsist on a less quantity
of food, or food of an inferior quality, may thereby
survive, but it does not follow that it will be improved
by the change ; on the contrary, it may be very much
the reverse, as the probability is that it will have
become less beautiful and smaller than it was before.
Sexual selection proceeds on different lines. Size,
strength, and beauty for themselves are what sexual
selection makes for. The female selects the handsomest
and most valiant male. Dull colours are affected by
natural selection, because they are useful ; bright colours
are preferred by sexual selection, because they are
beautiful, utility being the criterion in the one case and
beauty in the other.

The facts here presented lead me to the conclusion
that the sexual struggle is not what Darwin represents
it to be—a struggle between individuals of the same sex,
namely, the males—but rather a struggle between the
opposite sexes ; that all male sexual characters, secondary
as well as primary, must necessarily have relation to
female sex, and consequently that such characters cannot
be transferred from the one sex to the other.

CHAPTER V.

WHAT is the relationship between plants and insects? The Darwinist believes it to be of a most amiable and intimate character, and mutually advantageous. He will by no means admit that it is a one-sided affair. According to him both are benefited, especially the plants. They are supposed to have formed themselves into a sort of mutual benefit society. The insects obtain nectar from the flowers, and the flowers get fertilized by the insects. Fertilization is brought about only in a casual sort of way, and quite unintentionally on the part of the insects, the latter not having the remotest idea of benefiting the plant. Nevertheless we are told the plant is benefited, and that is the main thing; or, rather, it is not the actually existing plant but the progeny of the plant, which is, or rather is to be, benefited; which shows what a disinterested creature the plant must be to give away its nectar and ask for nothing in return, but goodnaturedly accept as a gift what was never intended for it, or even for its unborn progeny, if it could be said there is any intention at all in the matter. Of course it would be rank heresy to believe that the insects had the best of the bargain.

Nevertheless the plants, we are informed, highly

appreciate the arrangement entered into with their
insectivorous friends, whom they endeavour to attract by
donning their brightest colours, and rendering themselves
beautiful, in order that the insects may be induced to
come and enjoy themselves.   Thus says Darwin :—
" Many flowers have been rendered conspicuous for the
sake of guiding insects to them." *   " We certainly owe
the beauty and odour of our flowers and the storage of a
large supply of honey to the existence of insects." †   We
are further informed that the plants exercise discrimination in offering their attractions.   " It may be admitted
as almost certain that some structures, such as a narrow,
elongated nectary, or a long tubular corolla, have been
developed in order that certain kinds of insects alone
should obtain the nectar." ‡

Nectar, it seems, is the great, but as we shall see
further on, not the sole attraction.   Nearly all flowers
contain nectar, and to obtain this, we are told, is the
main object of the insects in visiting them, and that this
nectar has been provided for the special purpose of
supplying their wants ; and in order that they may have
no difficulty in finding their way to the nectary "not
only do the bright colours of the flowers serve to attract
insects, but dark-coloured streaks or marks are often
present, which Sprengel long ago maintained served as

* *Cross and Self-Fertilization of Plants and Insects*, 2nd ed., 1878,
p. 386.

† *Ibid.*, p. 383.   ‡ *Ibid.*, p. 384.

guides to the nectary." * We may here remark that the
conclusion is not strictly logical, as these marks are
found on many flowers which have no nectar at all.
We are also informed that many flowers are rendered
odoriferous for the purpose of attracting certain kinds of
insects; while other flowers "emit their odour chiefly,
or exclusively, in the evening in order that they may
not be visited by ill-adapted diurnal insects." † Why
diurnal insects should be objected to as "ill-adapted"
is somewhat puzzling, especially when no explanation
has been supplied.

But we have not yet exhausted the list of pleasing
attractions which the flowers put forth in order to
ensnare their insectivorous friends. Having discovered
that not only the flowers but also the leaves of certain
plants secrete nectar, and that this outside deposit, as
well as that inside the nectary, was sought after by
insects, a difficulty arose as to the connection of this
peculiarity with the fertilization of the flower. What
purpose could be served by the nectar on the leaves?
In searching for honey stored in the nectary, which is in
the interior base of the flower, insects might, perchance,
get dusted over with pollen, which they might carry to
other flowers and thus fertilize them; but how could
this result be brought about if the insects never entered
the flower? The question was a difficult one to answer;

* *Cross and Self-Fertilization of Plants and Insects*, p. 373.
† *Ibid.*, p. 375.

nevertheless the reply comes promptly and without a shadow of hesitancy. The nectar in this instance served a different purpose from that stored in the nectary. It had nothing to do with floral fertilization but was intended for purposes of defence ! " In some cases," Darwin tells us, "the secretion seems to attract insects as defenders of the plants, and may have been developed to a high degree for this special purpose, I have not the least doubt." * Strange to say this view is entertained by such writers as Delphino and Kerner, as well as by Darwinists of all shades of opinion.

Perhaps the most extraordinary instance of " purpose " on record is that given by Darwin on the authority of the late Dr. Crüger, director of the Botanical Gardens at Trinidad. The *Coryanthes* and an allied species of orchid provide a bucket, and fill it with a limpid and slightly sweet fluid, not for the insects to drink, but for them to disport in ; and it is necessary that the insects should be *twice* immersed in this fluid, and that they should *twice* emerge from it again, and then crawl along a certain track, before fertilization can be accomplished. On this point we must again quote Darwin's own words, which will be intelligible without the accompanying woodcuts to which he makes reference :—

The flowers (of *coryanthes*) are very large, and hang downwards. The distal portion of the labellum is converted into a

* *Cross and Self-Fertilization of Plants and Insects*, 2nd ed., p. 406.

large bucket. Two appendages arising from the narrowed base of the labellum stand directly over the bucket, and secrete so much fluid that drops may be seen falling into it. This fluid is limpid, and so slightly sweet that it does not deserve to be called nectar, though evidently of the same nature ; nor does it serve to attract insects. When the bucket is full the fluid overflows by the spout. This spout is closely overarched by the end of the column which bears the stigma and pollen masses in such a position that an insect forcing its way out of the bucket through this passage would first brush with its back against the stigma, and afterwards against the viscid discs of the pollinia, and thus remove them. We are now (he goes on to state) prepared to hear what Dr. Crüger says about the fertilization of an allied species, the *C. macrantha*, the labellum of which is provided with crests. I may premise that he sent me (Darwin continues) specimens of the bees which he saw gnawing these crests, and they belong, as I am informed by M. F. Smith, to the genus englossa. Dr. Crüger states that "these bees may be seen in great numbers disputing with each other for a place on the edge of the hypochial (*i.e.*, the basal part of the labellum). Partly by this contest, partly, perhaps, intoxicated by the matter they are indulging in, they tumble down into the bucket, half-full of a fluid. They then crawl along in the water towards the anterior side of the bucket, where there is a passage for them between the opening of this and the column. If one is early on the look-out, as these *Hymenopterœ* are early risers, one can see in every flower how fecundation is performed. The humble-bee, in forcing its way out of its involuntary bath, has to exert itself considerably, as the mouth of the opichil (*i.e.*, the distal part of the labellum) and the face of the column fit together exactly, and are very stiff and elastic. The first bee, then, which is immersed will have the gland of the pollen mass glued to its back. The insect then generally gets through the passage, and comes out with this peculiar appendage, to return nearly immediately to its feast, when it is generally precipitated a second time into the bucket, passing through the same opening, and so inserting the pollen masses into the stigma while it forces its way out, and thereby impregnating the same or some other flower. I have often seen this ; and sometimes there are so many of these humble-bees assembled that there is a continual procession of them through the passage specified." There cannot be the least doubt that the fertilization of the flower absolutely depends on insects crawling through the passage formed by the extremity of the labellum and the overarching column. If the large distal portion of the labellum or bucket had been dry, the bees could easily have escaped by flying away. Therefore, we must believe that the fluid is secreted by the appendages in such an extraordinary quantity, and is collected in the bucket, not as a palatable attraction for the bees, as these are known to gnaw the labellum, but for the sake of wetting their wings, and thus compelling them to crawl out through the passage.*

I do not for a moment dispute the facts here narrated. That the structure of plants is as described, that the bees act in the manner stated, and that the plants are fertilized by the bees in this extraordinary fashion I have no reason to doubt : but that the whole of this complicated apparatus of fluid-secreting glands, appendages, baths, and passages has been provided for the purpose of securing fertilization by insects is simply incredible. It is far more probable that the insects made use of the existing apparatus than that it had been expressly provided for them in order to get the alleged purpose effected.

It will be observed that Darwin uses the term "purpose" when describing the relations between plants and insects. It is necessary, therefore, to explain what he

---

* *The Fertilization of Orchids*, p. 173-4, 2nd ed., 1885.

means by " purpose." He does not use the term in the sense ordinarily understood by it. Purpose implies forethought, deliberation, and intention; it is the conception of an intelligent agent. But Darwin uses the term in an altogether different sense. His theory demands, in the first instance, two formidable conditions. The first is unlimited variations; the second, unlimited time; and to these he adds, as we have seen, " a capacity for change " on the part of the organism. Thus, if among billions of variations a plant happily produced a flower with a plumose stigma, the flower would be what is called anemophilous—that is, it would be capable of being fertilized by the wind; and having acquired this, let us say, advantageous form, it would leave behind it a larger progeny than other plants not so well endowed. Suppose, again, that among innumerable plants, in some far remote time, one of them happened to secrete honey in the interior base of the flower which attracted the visits of insects (supposing them to have been in existence at this period), a race of plants bearing entomophilous flowers—that is, flowers capable of being fertilized by insects—would be the result. In all this there is no foresight, deliberation, or intention, nothing but mere chance. The use of the term " purpose " in the Darwinian sense is, therefore, incorrect and misleading. Small wonder that ordinary readers are puzzled when they meet with the term so often in the literature of this school. Even Lange was misled by the term, as in describing the process of natural selection he compares it to the sportsman who, *in order* to shoot a hare, goes out into the fields and fires at random in every direction, thus implying a purpose or object, but showing utter want of skill or intelligence on the part of the agent, which was what Lange wished to point out by his illustration. No doubt intelligence is a superfluity in the Darwinian universe.

From the extracts given above, it appears beyond doubt that Darwin maintained that certain species of plants have acquired a certain structure, and secrete a certain product for the purpose of attracting certain species of insects. I do not dispute the fact of the structure, or of the product, but I take leave to doubt whether the structure has been acquired or the product secreted for the purpose alleged. Further on I shall point out that the structure has in some instances been misunderstood; and that in other instances both the structure and the presence of the nectar are capable of a different interpretation from what has been put upon them. That certain insects visit certain flowers and in penetrating to or returning from the nectary they get their bodies dusted over with pollen, a portion of which may get carried to other flowers of the same species, and that in this manner the ovules of one plant may be fertilized with the pollen from another, or what is called cross-fertilization takes place, no one nowadays disputes. But what is not admitted, though the Darwinist insists on it, and what has to be proved by better evidence than has yet been adduced is—(1) that certain flowers produce nectar for the purpose of attracting insects ; (2) that certain flowers have acquired a certain structure, and that certain other flowers, called dichogamous, have their pollen and stigma matured at different times in order to ensure cross-fertilization ; and (3) that cross-fertilization is necessary or beneficial.

1. That flowers produce nectar for the purpose of attracting insects is very far from being proved. Were insects attracted solely by the presence of the nectar, the fact might be accounted for in the manner stated ; but the nectar is not the sole attraction ; and here the Darwinist does not deal quite fairly with his facts. It is well known that the same kind of insects which eat the nectar also devour the pollen. The pollen must therefore be an attraction as well as the nectar. How is it, then, that the Darwinist, while so eloquent about the attraction of the nectar, is so reticent about the pollen ? If insects are attracted by the pollen as well as by the nectar—and Darwin himself admits it—and if the nectar is produced for the purpose of attracting insects, why does he not acknowledge that the pollen might also be provided for the same purpose ? The reason is obvious. Darwin believes that the nectar is provided for the insects because he has not yet discovered any other use for it in the economy of the plant ; but he cannot possibly believe that the pollen has been produced for the purpose

of being devoured by insects, if he keeps in mind the obvious fact, which he so constantly enforces and so often forgets, that " the production of seed is the chief end of the act of fertilization." *

Without the nectar no insect visitations, according to the Darwinist, and without the insects no cross-fertilization ; so that the nectar is the all-important thing for him.   It is a fact, nevertheless, that many species of flowers have no nectary, and produce no honey, and are yet constantly visited by insects.   Thus, *Lupinus luteus,* the poppy, the common broom, the potato, the sweet pea, and numerous other species, have no nectary, and secrete no honey, and yet are constantly visited by both hive and humble bees.   On the other hand, many species of plants, as *Epipactus latifolia,* have a copious supply of honey in the nectary which is never touched by insects.   In such cases, therefore, it cannot be said that the honey is the attraction.   The fact of the matter is that in these, and in most other cases, the insects are not allured by the nectar at all but by the pollen.   Darwin is of opinion that insects were originally attracted by the pollen, and subsequently by the nectar.   This may or may not have been the case ; but whether or not this much is certain, that if they commenced by eating the pollen and afterwards acquired a preference for the nectar, they have not yet lost their liking for pollen.

2. The argument from structure—the general form of

* *Cross and Self-Fertilization of Plants,* p. 3.

the flower, the position of the anthers and stigma, and the existence of dichogamous flowers—seems, at first sight, a strong one. Yet many species of plants, which appear to be specially adapted for cross-fertilization, are self-fertile, although constantly visited by insects. Thus, the sweet pea (*Lathyrus odoratus*) and the common pea (*Pisum sativum*) are constantly visited by insects, yet they are never fertilized by them. Of the latter, Mr. Farrer says that its structure peculiarly fits it for cross-fertilization—"The open blossom, displaying itself in the most attractive and convenient position for insects; the conspicuous vexillum, the wings forming an alighting place; the attachment of the wings to the keel, by which any body pressing on the former must press down the latter; the staminal tube, enclosing nectar, and affording, by means of its partially free stamen, with apertures on each side of its base, an open passage to an insect seeking the nectar; the moist and sticky pollen, placed just where it will be swept out of the apex of the keel against the entering insect; the stiff elastic style, so placed that on a pressure being applied to the keel it will be pushed upwards out of the keel; the hairs on the style placed on that side of the style only on which there is space for the pollen, and in such a direction as to sweep it out; and the stigma, so placed as to meet an entering insect—all these become correlated parts of one elaborate mechanism, if we suppose that the fertilization of these flowers is effected by the carriage of pollen from

one to the other."* Notwithstanding these manifest
provisions for cross-fertilization, Darwin has to admit
that varieties of this plant, which have been cultivated
for many successive generations in close proximity,
although flowering at the same time, continued to
remain pure.

Indeed, the argument from structure may be used on
the other side of the question, and with much greater
force. While the Darwinist is perpetually expatiating
on the wonderful contrivances for ensuring cross-fertilization, he carefully ignores the facts which point
unmistakably in an opposite direction. He omits, for
instance, to inform us that there is a larger number of
species which are more adapted for self-fertilization than
for cross-fertilization by insects, while many species have
their flowers so formed that cross-fertilization is an
absolute impossibility. Thus, hermaphrodite plants have
the anthers and stigma on the same flower, and these are
often so situated that it is almost impossible for self-
fertilization not to take place. There are numerous
species with small and inconspicuous flowers, which are
seldom or never visited by insects; and may we not
infer that if flowers have acquired size and conspicuousness in order to attract insects, that small and inconspicuous flowers have been so constructed to deter
insects from visiting them? Then there are other
species which have neither small nor inconspicuous

* *Nature*, 10th October, 1872, p. 479.

flowers, but have a narrow elongated nectary, or a long tubular corolla, sometimes with hairs inside, which prevents the admission of any except the smallest, and from the Darwinist point of view, the most useless kinds of insects. And, lastly, we have plants with cleistogamic or closed flowers, which no insect whatever can penetrate. Why does the Darwinist omit mention of such structures as these? There can be no mistake about their bearing on the question before us. If some flowers have, or appear to have, a structure adapted for cross-fertilization, here we have others which are unmistakably adapted for self-fertilization, and others again which it is impossible that insects can fertilize under any circumstances. What is the inference to be drawn from these facts? In the one case the adaptations must be regarded as imaginary, as we have seen it to be with the sweet pea and the common pea; in the other cases the adaptations are real because they are effective, cross-fertilization being rendered impossible. No one who has carefully observed the reproductive organs of plants can fail to have been struck by the overwhelming amount of evidence in favour of structural adaptation for the purpose of ensuring self-fertilization.*

The existence of dichogamy in plants no doubt favours cross-fertilization. Dichogamous plants are of two kinds,

---

* Darwin admits that "the greater number of flowers present structures which are manifestly adapted for self-fertilization."— *Cross and Self-Fertilization of Plants*, p. 381.

those, namely, in which the pollen is matured before the stigma, called proterandrous; and those in which the stigma is matured before the pollen, called proterogynous, the latter form being not nearly so common as the former. The question which concerns us here is, whether dichogamy has been acquired in order to ensure cross-fertilization by insects, or has been the result of insect visitation. I am of opinion that the latter view is the correct one. A plant accustomed to be fertilized by outside agency would not make any effort, and would not require to make any effort, to fertilize itself, and the mechanism would, consequently, no longer be adjusted to its proper functions, and the flower would adapt itself to its new conditions of existence. Most open flowers are adapted for either self or cross-fertilization, and circumstances will determine the action of the plant. If it has been in the habit of fertilizing itself there will be no necessity for insect visitation; on the other hand, if it has been accustomed to be fertilized by insects, insect visitation will be necessary, or sterility will be the result. Thus, *Mimulus luteus* fertilizes itself by bending down the pistil to the stamen; but if visited by insects the plant would be spared the effort of fertilizing itself, and would in course of time cease to make it. *Ipomea purpurea*, when full grown, fertilizes itself by brushing its anthers against the stigma; *Viola tricolor* by curling its petals inwards when the flower is mature; *Lobelia ramosa* by pushing the pollen out of the conjoined anthers; buttercups by closing themselves at night, thus bringing the stigma in contact with the anthers. But if flowers are habitually fertilized by outside agency they will lose their functional activity in this respect, their organic equilibrium will be disturbed, and in order to adapt themselves to their changed conditions, they will acquire certain modifications in their structure which will sometimes assume, as in the case of orchids, the most curious and even fantastical shapes.

And here comes into operation the law of Use and Disuse, or, as I should prefer to call it, of Effort and Abstinence, and of the inheritance of functionally produced modifications. This law Darwin insists on as firmly as does Mr. Herbert Spencer, notwithstanding its apparent contradiction to the theory of natural selection. "It is probable," he says, "that disuse has been the main agent in rendering organs rudimentary. On the whole we may conclude that habit, or use and disuse, have in some cases played a considerable part in the modification of the constitution and structure."* If disuse has been the main agent in rendering organs rudimentary or abortive, there can be little doubt of its application to the case before us. If from disuse our domestic ducks and geese have almost lost their power of flight; if with the domestic pigeon the length of the sternum, the prominence of its crest the length of the scapulæ and the furcula, and the length of the wings are all reduced

* *Cross and Self-Fertilization of Plants*, p. 114.

relatively to the same parts in the wild pigeon; if the shortened legs and snout of the domestic pig and the reduced size of the lungs and liver of our domestic cattle are due to disuse, as Darwin maintains—I say if all these modifications have been so caused, it is easy to comprehend how it is that disused functions in the sexual organs of plants may produce partial sterility and correlative modifications of structure.

On this principle we may account for proterandry and proterogyny. The reproductive organs being most sensitive, the constant visits of insects, especially of the larger species, will cause the parts of the flower on which they alight, and crawl over on their way to the nectary to mature early.* This also explains how it is that proterandry is much more common than proterogyny. As a rule the anthers are about five to one of the stigmas, and the former being more in the way of the insects when forcing their entrance into the nectary than the latter, which are in the centre, the former would be matured earlier than the latter. That plants have become proterandrous or proterogynous in order that they may not be self-fertilized seems to me most improbable.

3. In order to ascertain whether self or cross-fertilization is most beneficial, Darwin commenced a series of experiments, which he carried on for several years, the

* Mr. Henslow has some remarks on this subject in his able and suggestive work on *Floral Structures*.

results of which he has embodied in his work on *Cross and Self-Fertilization of Plants.* Since the publication of that book other botanists, both in England and Germany, have entered the same field of inquiry, but as they have all taken the same view as Darwin, and have not added anything material to our knowledge on this subject, I shall confine myself to an examination of the results given in the work referred to. The general conclusion which Darwin arrived at is this, that " cross-fertilization is generally beneficial, and self-fertilization often injurious ; " but he adds that " the most important conclusion at which I have arrived is, that the mere act of crossing by itself does no good. The good depends on the individuals which are crossed differing slightly in constitution, owing to their progenitors having been subjected, during several generations, to slightly different conditions, or what we call in our ignorance spontaneous variation." * Darwin conducted his experiments in the following manner :—He sowed the seed of various species of plants, taken respectively from flowers which had been cross-fertilized, and from others which had been self-fertilized, and these he grew under the same conditions ; when they were fully grown he measured their height, counted the seeds which they had produced, and in some instances cut down and weighed the plants.

Darwin's experiments, however, have been seriously

* *Cross and Self-Fertilization of Plants*, p. 27.

vitiated—(1.) By his selecting for experiment almost
exclusively plants which had long been cultivated in
our gardens and nurseries, which had large open flowers
which had been habitually fertilized by insects. It was
to be expected that a plant which had been accustomed
to be fertilized by insects would react very differently
towards its own pollen from what it would towards
pollen brought to it from another plant by an agency
outside itself.   (2). By cross-fertilizing by hand, instead
of allowing fertilization to take place by insects in a
natural way.   The difference between the two processes
is material. In cross-fertilizing by hand the operator
makes sure that the pollen is taken from a distinct plant,
probably from a fresh stock altogether ; but there is no
such guarantee if fertilization is left to insects. In the
latter case the pollen is almost sure to be gathered from
the nearest plant, or from the same plant.   It has been
observed that insects, and especially bees, which are the
chief agents in this process, almost invariably visit all
the flowers of the same species which are adjacent, one
after the other, before going to others of the same species
at a distance.   By this means the last-visited flower
would, almost to a certainty, receive the pollen from the
same or some adjacent plant.   (3.) By not extending his
experiments with individual species for longer periods.
As a rule he was content with one trial only, or one
generation of plants. When he departed from this rule
the results were sometimes remarkably different.

A few illustrations will show the importance of this last objection. Thus, with *Ipomea purpurea*, a self-fertile plant, but which is often crossed by insects, Darwin experimented for as many as ten successive generations, and in each generation, from the first to the tenth inclusive, the seedlings from the crossed plants in each trial were on an average taller than those from the self-fertilized plants of the same stock. But the difference in growth between the intercrossed and self-fertilized seedlings did not continue to increase with each successive generation, as we should expect if self-fertilization were injurious and cross-fertilization favourable. For the difference between the two sets of plants in the seventh and eighth generation was actually less than it was in the first and second. But although the seedlings from the intercrossed were, on an average, superior in height to those from the self-fertilized plants, a remarkably vigorous plant appeared among the latter in the sixth generation, which Darwin named " Hero." He was so surprised at the appearance of this plant that he resolved to ascertain whether it would transmit its power of growth to its progeny. Several flowers on "Hero" were accordingly fertilized with their own pollen, and the seedlings thus raised were put into competition with seedlings from intercrossed plants of the corresponding generation. The result proved, much to Darwin's astonishment, the great superiority of the seedlings from the self-fertilized "Hero" over the seedlings

from the intercrossed plants ; and, what seemed to the
experimenter still more extraordinary, all succeeding
self-fertilized descendants of "Hero," to the number of
seven generations, continued to show their superiority
when put into competition with seedlings from intercrossed plants of the corresponding generation. Thus
Darwin was forced to admit not only that the descendants
from "Hero" had inherited " a power of growth equal to
that of the ordinary intercrossed plants," but also that
they had " become more fertile when self-fertilized than
is usual with plants of the present species."*

The only other species which Darwin experimented on
for a number of generations was *Mimulus luteus*, a
species which is said to bear flowers " peculiarly
adapted for insect fertilization," and a result similar
to the last followed, only more decisive. In the first
four generations the seedlings from the cross-fertilized
were superior in height to the seedlings from the
self-fertilized plants ; but in the fifth generation the
position was reversed, and the latter gained the
superiority, not in height only, but also in fertility.
In the fifth generation the crossed plants were in
height to the self-fertilized as 100 to 126, in the sixth
as 100 to 147, in the seventh as 100 to 137; and "in the
sixth generation the self-fertilized plants of this variety
compared with the crossed plants produced capsules
in the proportion of 147 to 100," and in the seventh

* *Cross and Self-Fertilization of Plants*, p. 50.

generation the self-fertilized plants were equally fertile.*

Even the results of experiments carried on for only one or two generations did not always prove the superiority of seedlings from crossed plants. Thus he experimented with *Eschscholtzia Californica* for two generations; in the first the seedlings from crossed exceeded in height those from self-fertilized plants as 100 to 86, but in the second generation the crossed stood to the self-fertilized as 100 to 101. But what is even more remarkable is that seedlings from plants crossed with fresh stock (which Darwin considers to be immensely beneficial), proved inferior to the seedlings from self-fertilized plants in the proportion of 100 to 116 in height and 100 to 118 in fertility. *Petunia violacea* is another case in point. In the first and second generations the seedlings from crossed were decidedly superior in height to those from self-fertilized plants, but in the third generation the latter beat the former in two trials, first when the plants were young, when the crossed were to the self-fertilized as 100 to 186, and next when they were fully grown, when they were as 100 to 131. Seedlings from these self-fertilized plants were next tested for fertility, when they proved superior to the crossed in this respect also, although in the two previous generations they had been inferior. Every breeder of domestic animals knows that a first cross is usually

* *Cross and Self-Fertilization of Plants*, p. 79.

beneficial, but that a second or subsequent crosses are almost certain to be the reverse.

On the whole it appears that in 36 out of 54 species the crossed exceeded the self-fertilized in height, while in the remaining 18 the position was reversed, the self-fertilized in these instances exceeding the crossed. As regards fertility, also, Darwin's experiments show that the cross-fertilized plants had the advantage.*

* In some respects, indeed, Darwin has not put the results of his experiments quite fairly. He recommends the general reader in the introduction to skip the details of his experiments and read his summaries and observations at the end of his work ; but I find the summaries at the end do not always agree with the details at the beginning. For instance, he made two experiments with *Eschscholtzia Californica* ; in the first the crossed were to the self-fertilized in height as 100 to 86, in the second as 100 to 101, and he puts the former in his summary and omits the latter. He made experiments with *Mimulus luteus* for several generations, but in his summary he only gives the results of the first three, which supported his views, and omitted all subsequent results, which were adverse. With *Digitalis purpurea* he made two experiments, in both of which the crossed plants had the advantage in height, but one more than the other, and he omits from his summary the one which was least favourable to his own views. Two experiments were made with *Brassica oleracea*, the same plants on each occasion, one before they were fully grown and the other after they had seeded. In the first trial the self-fertilized exceeded the crossed in height, in the second the result was reversed ; he omits the first measurement altogether, and gives only the last. In two experiments with two different sets of plants of *Viscaria osculata*, the first showed the crossed to be inferior, and it is omitted ; the second showed the reverse, and it is given. With *Lobelia fulgens* he made two experiments on the same sets of plants.

Darwin's experiments show one very important result, which has not received at his hands that attention which it deserves. I refer to the fact that continuous self-fertilization does not produce deterioration. The experiments, so far as they go, show not only that the latest generations of plants, whose ancestors had long been self-fertilized, suffered no injury from the process, and also that cross-fertilized plants, whose ancestors had long been crossed, received no benefit in the later generations. *Ipomea purpurea* and *Mimulus luteus*, as already stated, were the only species which were experimented on for long periods, and the self-fertilized plants of the former species showed no deterioration in height, as

In the first the crossed were to the self-fertilized as 100 to 127, and being, as he says, much surprised at this result, he determined to try how they would behave during a second growth. In this second experiment the crossed were to the self-fertilized as 100 to 167. The results of the first experiment are given and those of the second omitted. In the case of *Brassica oleracea*, given above, it will be observed that he adopted exactly the opposite course. Three experiments were made with *Nemophila insignis*, the two first on the same sets of plants at different stages of growth, in which the crossed were to the self-fertilized as 100 to 49 and 100 to 60 respectively; in the third experiment the crossed were to the self-fertilized as 100 to 133. The results of the two first are given and the third omitted. With *Petunia violacea* of the third generation he made two experiments with the same sets of plants, the first when young, when the crossed were to the self-fertilized as 100 to 186; the second when fully grown, when the proportion was reduced to 100 and 131 respectively, and the latter measurement only is given.

8

compared with the crossed plants of the corresponding generation—in fact, rather a steady improvement—and very little in fertility ; while the self-fertilized plants of the latter species compared with the crossed showed a marked and steady improvement from the first to the tenth generation, both in height and in fertility. Now, if self-fertilization is injurious, each generation of self-fertilized plants should have been worse than the previous one, and each generation of the crossed should have been better.

The result of his experiments Darwin sums up in the statement that " the mere act of crossing by itself does no good ; " the good, he says, depends on the individuals which are crossed differing in constitution, from being grown under different conditions. A cross with individuals growing for any length of time on the same soil will therefore be of little or no advantage. The reason is obvious. Plants, when they remain for a long time in the same place, exhaust the soil of certain constituents which are necessary for their well-being, and the longer they remain there they will become less and less vigorous, and less and less fertile, and, in course of time, even altogether sterile, for sterility is one of the results of low vitality. The deterioration of the plants is entirely due to the exhaustion of the soil. Change the soil, or remove the plants to another situation where the soil has not been exhausted, and the plants will at once recover their wonted vigour and fertility. But no permanently good effect will be obtained by crossing plants which are suffering from want of proper sustenance, not even by crossing with a fresh stock. At best a cross will only produce a temporary improvement under such circumstances.

For this state of things nature has provided a remedy, not by transporting the pollen from one flower to another, but by transporting the seed to fresh soil. For this purpose many kinds of seeds are of extraordinary lightness, and can be carried by the wind great distances ; some kinds are provided with plumes and wings for floating in the air; others have hooks and spears which fasten on, or penetrate into, the bodies of animals and are thus carried far and wide ; while others, again, have capsules with peculiar mechanical contrivances, which explode when the seed is ripe, and scatter it to a distance from the plant. The seeds of many water plants are also endowed with locomotive power, by means of vibratile hair-like processes, called cilia. Similar provisions have been made in the animal world for the distribution of seed. All stationary marine and freshwater animals produce young, which, by means of flagella and cilia, have the power of moving about freely. The air, the earth, and the water are full of the germs of vegetable and animal life. If a handful of soil be exposed to sun and air a number of plants will immediately make their appearance, many of which belong to distant localities, the seed having been transported hither by the agencies referred to. That the air we breathe is full of the germs of organisms, both animal and vegetable, has been amply proved by the researches of Pasteur, Frankland, and others; and not near the surface only, but at great elevations; and at sea, more than a hundred miles from land.

There is no evidence whatever of want of vigour or of sterility in plants which either have not been or cannot be fertilized by insects. On the contrary the evidence is all the other way. Darwin admits that some plants, owing to their structure, have almost certainly been propagated in a state of nature for thousands of generations without having been once intercrossed. * Plants with cleistogamic or inconspicuous flowers are, of course, self-fertile; and if self-fertilization were injurious we should look for some evidence of it in this class of plants. We should expect them to be weak and unfertile, and consequently rare. But the contrary is the case. As a rule such plants hold their own against all competitors, and are vigorous and prolific to an extraordinary degree. They seem to flourish everywhere— no temperature being too hot or too cold, no soil too poor or too rich for them. The great majority of weeds belong to this class. *Malva rotundifolia* has established itself in every part of the world; *Stellaria media*, or the common chickweed, is to be found not only all over Europe, but in Australia, New Zealand, Tasmania, South

* *Cross and Self-Fertilization of Plants*, p. 442.

Africa, North and South America, India, and the Auckland Islands.

There can be little doubt, I think, that cultivated plants, like domesticated animals, are more subject to variation than plants in a state of nature. Almost all our domestic animals (the goose and the guinea-fowl being almost the only exceptions)—rabbits, fowls, pigeons, and cattle—have varied extensively both in form and colour; and it is a singular fact that all of them when allowed their freedom revert to their aboriginal type. Although it has not been observed that plants show this tendency, it is well known that in their wild state their colours seldom vary. It is also a fact, which is amply confirmed by Darwin's experiments, that cultivated plants, which vary excessively in colour when crossed, lose this tendency when self-fertilized for a number of generations, and become absolutely uniform.

With most plants which have been long cultivated for the flower garden, no character is more variable than that of colour, excepting perhaps that of height. . . . The variability of plants in the above respects is due, firstly, to their being subjected to somewhat diversified conditions, and secondly, to their being often intercrossed, as would follow from the free access of insects. I do not see how this inference can be avoided, as when the above plants were cultivated for several generations under closely similar conditions, and were intercrossed in each generation, the colour of their flowers tended in some degree to change and become uniform. When no intercrossing with other plants of the same stock was allowed—that is, when the flowers were fertilized with their own pollen in each generation—their colour in the latter generations became as uniform as that of plants growing in a state of nature.[*]

No doubt self-fertilization is a great factor in producing uniformity of colour. That this uniformity is not due to the plants having been " subjected to somewhat diversified conditions," as Darwin intimates, is shown by the fact that wild plants grow and maintain their uniform colour in all sorts of soils and in all kinds of situations. But if self-fertilization be the cause of uniformity, the converse will also hold good, and variation must be regarded as the result of cross-fertilization. This may account for the excessive variations alike in domesticated animals and cultivated plants. We know that with regard to animals in a state of nature consanguinity is no bar to marriage. Here the most closely related animals pair with one another; and among polygamous animals the male mates freely with his own progeny and drives away or kills every strange male that ventures near his females. With domestic animals it is different. Here in-breeding is the exception; and it is not till he has produced the type of animal that he wishes to propagate that the breeder permits in-breeding, and this he does ultimately in order to establish his type. Now, to what do these facts point ? Darwin insists upon it that cross-fertilization is beneficial; but if we take into

[*] *Cross and Self-Fertilization of Plants*, pp. 310-311.

account the indisputable facts as to the uniformity of
colour of plants in a state of nature, and their diversity
under cultivation, and if we further admit that this
uniformity and this diversity are brought about in the
manner indicated, I do not see how we can avoid the
conclusion that natural selection has failed, in the case
of wild plants, to appropriate the advantage offered to it
by cross-fertilization. Either we must accept this conclusion or refuse to admit that cross-fertilization is
beneficial.

Assuming, however, that cross-fertilization is beneficial, and that it is brought about by means of insects,
we should expect, in accordance with the principle of
natural selection, to find that wherever honey-eating
and pollen-devouring insects were abundant, there the
flowers would be large and conspicuous, and, as a consequence, cross-fertilization would take place to a large
extent, and the progeny from the intercrossed plants
would be so vigorous and so fertile that they would in the
course of time outnumber, if they did not take the place
of, self-fertilized plants. In other words, we should have
entomophilous flowers everywhere prevalent. But as a
matter of fact we find no such results. Entomophilous
flowers, whether reckoned by species or by individuals,
are by no means numerous, compared with other flowers
which are not visited by insects. They are common enough
among our cultivated plants, it is true, and that is how we
come to know them so well ; but they are comparatively

rare in a state of nature. The list of non-entomophilous
plants, on the other hand, is extremely large, including
not only all those classed as hermaphrodite, many of
which are also fertilized by insects, amenophilous (which
in the number of individuals, if not in species, alone
exceeds the entomophilous plants), the cleistogamic, and
such as bear small and inconspicuous flowers, but also
whole groups of species and genera which do not produce
seed at all, and are, therefore, incapable of being intercrossed, but propagate by means of rhizomes, soboles,
corms, cloves, and bulbs. There is no special virtue in
seed any more than there is in a cross. Many species of
plants produce both seeds and bulbs, and propagate more
freely by the latter than by the former, and yet show no
loss of vigour or fertility—as, for instance, potatoes, and
tubers of almost every description. Our common fruit
trees have also been propagated for thousands of years
by cuttings, and yet we have no reason to believe that
they are not as fruitful as ever they were. The grape
vine is even a better illustration, for it has been propagated by cuttings from time immemorial, and as far
back as we can trace it has been subjected to the hardest
usage by severe pruning, and yet it gives no sign of
deterioration.

But I shall probably be told that the process of con-
version of plants from non-entomophilous to entomophilous
is still going on, and that in course of time the latter
will possess the earth. True, I forgot that the Darwinist

demands immeasurable time in the future as well as in the past. It is difficult to deal with questions involving what are practically infinite quantities, but even an immeasurable future will not serve him in the present instance, as entomophilous plants never can be very numerous, for many reasons, for—1. They can only advance in masses; all stragglers from the main body will die and have no progeny, as they cannot be fertilized in the absence of other plants of the same species. 2. The masses are liable at any time to die without progeny if, from any cause, there should be no insects to fertilize them. 3. Throughout a great portion of the globe, as in the Arctic and Antarctic and the colder temperate and alpine regions, insects are either comparatively rare or altogether absent. 4. Even where insects are abundant, as in the tropics, insect fertilization does not invariably take place. Darwin himself admits that fertilization of insects, even under the most favourable conditions, is often a failure. Thus he says :—

The frequency with which throughout the world members of various orchideous tribes fail to have their flowers fertilized, though these are excellently constructed for cross-fertilization, is a remarkable fact. Fritz Müller informs me (he goes on to say) that this holds good in the luxuriant forests of South Brazil with most of the *Epidendriæ* and with the genus *Vanilla*. For instance, he visited a site where vanilla creeps over almost every tree, and although the plants have been covered with flowers, yet only two seed capsules were produced. So, again, with *Epidendrum*, 233 flowers had fallen off unimpregnated, and only one capsule had been formed; of the still remaining 136 flowers only

four had their pollinia removed. In New South Wales, Mr. Fitz-
gerald does not believe that more than one flower out of a thousand
of *Dendrobium speciosum* set a capsule, and some other species there
are very sterile. In New Zealand over 200 flowers of *Coryanthes
trilobia* yielded only five capsules, and at the Cape of Good Hope
only the same number were produced by 78 flowers of *Disa grandi-
flora*. Nearly the same result has been observed with some of the
species of ophrys in Europe. The sterility in these cases is very
difficult to explain. It manifestly depends on the flowers being
constructed with such elaborate care for cross-fertilization that
they cannot yield seed without the aid of insects. . . . In
these cases in which only a few flowers are impregnated, owing
to the proper insects visiting only a few, this may be a great
injury to the plant, and many hundred species throughout the
world have been thus exterminated.[*]

We should also expect to find the number of entomophilous and bright-coloured flowers to be in proportion to the number of nectar and pollen eating insects.
As a matter of fact we find that wherever insects are
most numerous, conspicuous flowers are rarest, and
wherever conspicuous flowers are most abundant, insects
are comparatively scarce, or altogether absent. Conspicuous flowers are comparatively rare in the tropics,
where insects are abundant, and they are numerous in
temperate zones, where insects are scarce. Drs. Wallace
and Spence, Mr. Bates, and others have expressed
their disappointment at the sparseness of the floral
display in tropical regions. Nowhere are brilliant
masses of colour to be met with, such as are of

[*] *Cross and Self-Fertilization of Plants*, pp. 280-82.

common occurrence in temperate or alpine regions. Dr. Wallace says:—

It is when we come to the vegetable world that the greatest misconception in this respect prevails. In the abundance and variety of colour the tropics are almost universally believed to be pre-eminent, not only absolutely, but relatively to the whole mass of vegetation and the total number of species. Twelve years of observation among the vegetation of the Eastern and Western tropics has, however, convinced me that this notion is entirely erroneous, and that, in proportion to the whole number of species of plants, those having gaily coloured flowers are actually more abundant in the temperate zones than between the tropics.[*]

Mr. Bates, speaking of the forests of the Lower Amazon to the same effect, asks—" But where were the flowers? To our great disappointment we saw none, or only such as were insignificant in appearance. Orchids are rare in the dense forests of the lowlands, and I believe it is now tolerably well ascertained that the majority of forest trees in equatorial Brazil have small and inconspicuous flowers."[†] Dr. Spence assured Dr. Wallace that by far the greater number of plants gathered by him in equatorial America had inconspicuous green or white flowers, and Dr. Wallace acknowledges that his own experience in the Aru Islands and in Borneo is quite in accordance with this view.[‡]

[*] *Tropical Nature*, p. 165.
[†] *The Naturalist on the River Amazon*, 2nd ed., p. 38.
[‡] *Tropical Nature*, p. 61.

Notwithstanding the paucity of conspicuous flowers in tropical regions, the abundance of insect life is remarkable. " Wherever in the equatorial zone," says Dr. Wallace, " a considerable extent of primeval forest remains, the observer can hardly fail to be struck by the abundance and the conspicuous beauty of the butterflies," * and he says the number of species "everywhere very much surpasses the number in the temperate zone." † Nor are other insects wanting. " The hymenopterous insects of the tropics are," he says, " next to the butterflies, those which come most prominently before the traveller, as they love the sunshine, frequent gardens, houses, and roadways, as well as the forest shades, never seek concealment, and are many of them remarkable for their size or form, or are adorned with beautiful colours and conspicuous markings;" ‡ elsewhere § he says of bees and wasps, which are supposed to be the chief agents in cross-fertilization, that they are "excessively numerous " in the tropics.

In the temperate and alpine regions just the opposite of all this occurs, the flowers being conspicuous and insects rare or altogether absent. " The beauty of alpine flowers," says Dr. Wallace, ‖ " is almost proverbial. It consists either in the increased size of individual flowers, as compared with the whole plant, in increased intensity of colour, or in the massing of small flowers in dense

* *Tropical Nature*, p. 73.   † *Ibid.*, p. 74.
‡ *Ibid.*, p. 80.   § *Ibid.*, p. 90.   ‖ *Ibid.*, p. 232.

cushions of bright colour; and it is only in the higher
Alps, above the limits of forests, and upwards towards
the perpetual snow line, that these characteristics are
fully exhibited." *   Dr. Wallace, singularly enough,
considers that this conspicuousness may be traced
" to the comparative scarcity of winged insects in
the higher regions." †   He further explains that not
only do the flowers in the alpine regions differ from
those on the low-lands, but even the leaves and
fruit. " We find," he says, " the yellow primrose of the
plains replaced by pink and magenta-coloured alpine
species ; the straggling wild pinks of the lowlands by
the masses of large flowers in such mountain species as
Dianthus alpinus and *D. glacialis;* the saxifrages of the
high Alps, with bunches of flowers a foot long, as in
Saxifraga longifolia and *S. cotyledon,* or forming masses
of flowers as *S. oppositifolia* ; while the soapworts,
silenes, and louseworts are equally superior to the allied
species of the plains." ‡   The further we advance towards
the north the more the leaves increase in size; the fruits
of cultivated plants acquire a deeper hue, and the green
colour of the leaves becomes more intense than on the
plains. It is evident that these changes are due to
climatic influences ; but whatever may be the correct
explanation of the phenomena, the facts we have
adduced ought to dispel the idea that conspicuous colours

---

* *Tropical Nature,* p. 232.
† *Ibid.,* p. 232.   ‡ *Ibid.,* p. 232.

have been acquired for the purpose of attracting the visits of insects.

There is another aspect in which this question may be viewed. If plants gain more than they lose by the visits of insects, we can understand how natural selection would come into operation. But suppose the reverse to be the case, what then? Darwin has distinctly laid down the principle that if it can be proved, by a single instance, that one organism exists for the benefit of another organism, his whole system would fall to the ground. The question is, Do the losses which plants sustain from insects outweigh the benefits derived from them, or do they not? Or, to put the question in another way, Are insects the friends or foes of plants? They are certainly not the friends, if we may judge of them by their acts. They destroy the bloom and scratch the petals of the flowers with their hooked tarsi; they eat the leaves and stems, and even the roots of plants; they bore holes in the stem and branches, and kill the plants by eating out their cores; they devour almost every kind of vegetable organism which comes in their way, and what they do not devour they deface or destroy. There is not a plant in existence which escapes their ravages, and scarcely a species which has not its special insect parasite. Insects are in general so utterly destructive of vegetable life that it is difficult to believe that anything but evil can result from their visits.

I shall of course be reminded that it is only certain

species of insects that are claimed as useful—those, namely, that are attracted to the flowers by the honey in the nectary. But the visits of even these are not altogether an unmixed good, for in order to get at the nectar they often do incalculable injury to the plant, even if we leave out of account the loss of the nectar, which we must assume to be in some way necessary to the maturing of the seed. The insects known to fertilize flowers are hive and humble bees, butterflies, moths, wasps, flies, and thrips. The three last are of comparatively little use in cross-fertilization; butterflies and moths are more effective in this respect, but many species can reach both the pollen and the nectar by means of their long proboscises without entering the flower, and of course without fertilizing it, and can therefore be of no service to the plant. Hive and humble bees are undoubtedly the chief agents in the distribution of the pollen; but in order that the pollen may be of use it must reach the stigma of the flower, and in order that it may reach the stigma the insects must enter the flower in the proper way. The nectar and the pollen are the attractions—the former being stored away in the nectary for the purpose, as we are told, of inducing the insects to crawl in and leave some pollen on the stigma and take away some fresh pollen from the anthers to distribute elsewhere. But unfortunately the nectar is placed in a somewhat inaccessible position, the silly plant having rather overshot the mark in placing it so much beyond the reach of the insects, so that humble bees (and hive-bees, too, if we are to believe Herman Müller and Darwin),* not being troubled with scruples of conscience, appropriate the nectar in a burglarious way. Although the door is open for them to walk in and help themselves in a proper manner, they frequently make an aperture in the calyx or the corolla, and in this way extract the honey from the outside. When they do this cross-fertilization cannot take place, and thus the very object for which the whole of the elaborate machinery was set in motion is defeated. It appears, also, that other insects, hive-bees in particular, finding a convenient entrance to the nectary, follow the example of the humble bees, and steal the honey through the aperture, and, of course, do not fertilize the flower any more than the humble bees.

The damage done in this way by humble bees is sometimes enormous. Darwin relates that he had seen whole fields of *Trifolium pratense* rendered infertile by the humble bees cutting the calyx of the flowers in order to get at the nectar,† and that in an extensive heath of *Erica tetralix* every flower he examined had been per-

---

* *Cross and Self-Fertilization of Plants*, pp. 430-435.

† Darwin says that *T. pratense* will not produce seed unless it has been visited by humble bees. The statement has been accepted without question, and some settlers in New Zealand have imported humble bees into that colony, in order to secure seed from the flowers, which bloom freely enough, but were believed, on Darwin's authority, to be infertile. But this is quite a mistake. Red clover seed had been grown and exported from New Zealand long before

forated by humble bees.* Mr. Belt states that he saw a late crop of *Phaseolus multiflorus* near London which had been rendered barren by humble bees cutting holes in the bases of the flowers. Humble bees have been seen to perforate both the calyx and the corolla of the same flower; indeed, they are not at all gentle creatures, for in their eagerness to get at the nectar they sometimes tear the flowers to pieces. The list of plants whose flowers they are known to perforate includes a great variety of species, too numerous to be given here; enough, I think, has been said to show that humble bees are of doubtful advantage to the plants they favour with their visits.

Hive-bees are, in some respects, even greater pests than humble bees, for they are robbers in a double sense; they not only eat the nectar but they also devour the pollen. Now, pollen is necessary for the production of seed, and the production of seed is the great end of the act of fertilization. Many insects, and especially hive-bees, are known to visit the flowers for the pollen alone, but the most pronounced advocate of insect fertilization will hardly contend that pollen was intended to be eaten Yet hive-bees eat it greedily, and in large quantities;

the humble bee was introduced there; and I am informed by one of the leading Melbourne seedsmen that he has been supplied with this seed, grown in the Western District of Victoria, for the last 17 years, although no humble bees have ever been introduced into that colony.

* *Cross and Self-Fertilization of Plants*, p. 429.

9

they also collect and store it for the purpose of feeding
their young. By means of the pencil of hair on their
tarsi they gather the pollen, which they knead into a
ball, and place it in the space situated in the joint or
tābia of the hinder leg, termed the basket, and in order
to carry away large quantities they roll their bodies on
the flowers, which they seriously damage by the process,
and then brush off the pollen which adheres to their
bodies with their feet—the very pollen which is supposed to be carried away to fertilize other flowers. It is
evident that the quantity of pollen which they carry on
their body bears a small proportion to what they eat or
destroy, while their incessant raids and their action in
rolling on the flowers can hardly be regarded as beneficial
to the plants. Darwin has noted the number of flowers
bees visit in a given time, and here is the result of his
observations :—

In the course of fifteen minutes a single flower on the summit
of a plant of *Oenothera* was visited eight times by humble bees,
and I followed the last of these bees whilst it visited, in the course
of a few additional minutes, every plant of the same species in a
large flower garden. In nineteen minutes every flower on a small
plant of *Nemophila insignis* was visited twice. In one minute six
flowers of *Campanula* were entered by a pollen-collecting hive-
bee ; and bees, when thus employed, work slower than when
sucking nectar. Lastly, several flower stalks on a plant of
Dictamnus fraxinella were observed on the 15th June, 1841 ; during
ten minutes they were visited by thirteen humble bees, each of
which entered many flowers. On the 22nd the same flower stalks
were visited within the same time by eleven humble bees. This

plant bore altogether 280 flowers, and from the above data, taking
into consideration how late in the evening humble bees work, each
flower must have been visited at least thirty times daily, and the
same flower keeps open during several days. The frequency of
the visits of bees is also sometimes shown by the manner in which
the petals are scratched by the hooked tarsi. I have seen large
beds of *Mimulus, Strachys,* and *Lathyrus* with the beauty of their
flowers thus sadly defaced.*

Just imagine the effect on a delicate flower of thirteen
visits from humble bees in ten minutes, or six visits a
minute from hive-bees! It gives one a pretty good idea
of the activity of bees, no doubt, but it suggests a conclusion which Darwin has omitted to draw. Are we to
believe that these hordes of intruders which deface the
beauty of the flowers in the manner here described are
friendly visitors?

I am aware that the view here presented does not
account for the beauty of flowers, which Darwin maintains we owe to the visits of insects. But neither does
Darwin's theory account for the beauty of open, self-
fertile flowers, nor for double flowers, which are infertile
and yet are more beautiful and conspicuous than single
flowers. Nor does it account for the beauty of the leaves,
the bright green in the spring, and the scarlet colours in
autumn, nor for the graceful forms of tree, shrub, and
flower.

Summary.—I cannot believe that flowers have ac-

* *Cross and Self-Fertilization of Plants,* p. 428.

quired their nectar, their present structure, and their
conspicuous colours in order to attract insects with a
view to cross-fertilization, as we have seen that pollen,
as well as nectar, is the attraction; that the adaptation
of structure to the visits of insects is often illusory, and
when real is not always effective for the purpose alleged;
that dichogamy is not produced in order to ensure cross-
fertilization, but is caused by the visits of insects. I
have endeavoured to show that Darwin's experiments to
test the effects of cross and self-fertilization were not
satisfactory, inasmuch as he selected the wrong kind of
plants, and did not test them for a sufficiently long
period; that, even with the plants selected, the results do
not show that cross-fertilization is beneficial; that there
is no virtue in a cross, and that even a cross with a fresh
stock produces only a temporary improvement; that the
offspring from parents which had been self-fertilized for
a number of generations did not show a gradual degeneration, nor the progeny from parents which had been
cross-fertilized for a lengthened period a gradual improvement; that on the whole the offspring of self-
fertilized plants maintain their vigour and fertility
equally with the progeny of intercrossed plants; that
many self-fertile plants were more vigorous and fertile
than many which were intercrossed, and that several
species, more particularly orchids, which appear to be
specially adapted for insect fertilization, are exceedingly
difficult to fertilize by insects, and are consequently

comparatively infertile. I have also shown that insects of all kinds are in various ways destructive to plants, and that even those species which assist in fertilization destroy the beauty of the flowers they visit, perforate holes in vital parts, and devour the pollen which they are supposed to use for fertilizing purposes. I also maintain that if insects were as beneficial to the plants as they are supposed to be, natural selection would have made use of them to a greater extent than appears to have been the case, and that entomophilous plants, instead of being less would have been more numerous both in species and individuals than all the other plants put together; and I argue that if, as Darwin contends, flowers owe their conspicuous colour to insects, we should have expected that wherever insects were most abundant flowers would be most conspicuous, and wherever insects were scarce that flowers would be inconspicuous, instead of the reverse, which we found to be the case. And, lastly, I point out that plants which depend on insects for fertilization can never become numerous, because they can only advance in masses; that masses are liable to die off in the absence of insects from any cause; that throughout a great portion of the globe, as in the arctic, antarctic, temperate, and alpine regions, insects are comparatively rare or altogether absent; and that even where they are abundant, as in the tropics, they fail in fertilizing the flowers, which are barren to a degree, and neither beautiful nor conspicuous.

CHAPTER VI.

THE CAUSES OF VARIABILITY.

PROFESSOR Huxley has somewhere said that what we want is a good theory of variability. There can be no doubt that a sound theory would dispel many existing illusions on the subject of natural selection. Generally speaking, the factors concerned in the production of variations are of two kinds—namely, internal and external, or the nature of the organism and the nature of the conditions under which it exists. It is not to be supposed that the organism, in the absence of an inciting cause, would spontaneously modify itself; and it would be absurd to imagine that the external conditions would directly produce variations without the co-operation of the organism. Both factors must concur in order to produce any effect.

That the external conditions, as food, climate, and situation, have great influence in modifying organisms is beyond a doubt. Plants and animals vary in a remarkable manner according to the quantity or quality of the food on which they subsist, the temperature in which they live, the atmosphere in which they breathe, and the dryness or dampness of the locality which they inhabit. But we have no precise knowledge as to the effects which any one of these various conditions, far less any combination of them, have in producing variability, as no experiments, on an adequate scale, have been carried out in order to test their individual or collective influence on living organisms. Of this much we may be certain, however, that if any organism be placed under conditions to which it has not been accustomed it will endeavour to adapt itself to these conditions, whatever they are, and a certain amount of variation will consequently take place; and if no variation occur we may conclude either that the external conditions were not uniform or powerful enough, or that they had not existed for a sufficiently long period to leave any permanent effect behind them.

Schmankewitsch's experiments on *Artimia* show the remarkable influence which external conditions exercise on organic life. There are two species of this crustacean, *Milhausanina* and *Salina*, the former a freshwater and the latter a saltwater species, and Schmankewitsch showed that it was possible to raise a brood of *A. Milhausenii* from *A. salina*, and, conversely, a brood of *A. salina* from *A. Milhausenii*, by gradually raising or lowering the percentage of salt in the water from a minimum of 4° B. to a maximum of 25° B. The transformation from the one form to the other was, however, very gradual, having taken several generations to effect. Schmankewitsch was able to carry his experiments to a

still further issue, for he kept gradually diluting the
salt water in which he kept *A. salina* till at last it was
perfectly fresh ; the crustaceans had in the meantime
gone through several generations and had so gradually
and completely changed their nature that they finally
acquired the characteristics of the genus *Branchipus.*

Semper's experiments on the common pond snail
(Lymnaea stagnalis) are also highly interesting. In
order to ascertain if the volume of water had any effect
on these animals, he instituted two series of experiments
—one by separating the animals from the same mass of
eggs immediately they were hatched, and placing them
simultaneously in unequal bodies of water; the other
by placing two different lots of animals, from the
same mass of eggs, in two aquaria of equal size. All
the conditions of existence, and above all the supply of
food, were kept at the optimum, consequently all the
animals were under equally favourable conditions, irre-
spective only of the volume of water which fell to each
animal's share ; this varied between 100 and 2,000 cubic
centimetres. In both experiments the results were similar ; the smaller the volume of water which fell to the
share of each animal, the shorter its shell remained. *

The effect of food in inducing variations may be observed in the case of cultivated plants, which, being well
nourished, show greater variability than those in a state
of nature. There is a similar tendency to vary among

* *Animal Life*, pp. 160, 161.

animals in the state of domestication, and from the same
cause, as witness the numerous varieties of domestic
pigeons, fowls, and dogs. So strong is this tendency to
vary with our domestic animals that scarcely two of a
litter or two of a brood are alike in colour, form, size, or
disposition. Facts such as these justify the conclusion
that the struggle for existence which takes place in a
state of nature prevents the free development of any
latent qualities in the organism which is subjected to
this process.

Situation also exercises a modifying influence. A plant
growing in a damp place usually has its leaves less
divided, is more glabrous, and has smaller and darker
coloured flowers than one grown on dry porous soil.
Thus in a damp soil *Dianthus alpinus* is transformed, in
the second generation, into *deltiodes*; in a dry and
porous soil *Hutchensia breircaulis* passes into *H. alpina*,
and *Arabis cærula* into *belliafolia*.

The intercrossing of varieties and of species is another
cause of variation. The offspring of hermaphrodite plants
seldom vary from the parent type. Plants propagated
from tubers, stolans, and buds also remain true to type,
and so also are the progeny of animals which propagate
by fision. Plants and trees grown from cuttings never
vary from the parent, no matter how long they may
have been cultivated: our fruit trees and vines have been
propagated in this manner for hundreds of generations,
yet they have never changed their character since the

first cutting was put into the ground. It is altogether different with diœcious plants, and animals which are propagated sexually ; these vary in a remarkable degree, so much so that it is seldom we find two plants or two animals exactly alike. Still they do not vary indefinitely. Half the flowers produced by seedlings from a plant which has been fertilized by another plant will probably be coloured like the one parent, and the other half like the other parent, or they may all have varying shades of colour intermediate between that of the two parents. Extreme crosses, as between well-marked varieties or between two species, will, however, prove a fertile cause of variation.

Plants and animals do not readily accommodate themselves to extreme changes of climate. Even seeds grown in a tropical and planted in a temperate climate rarely do well in their new situation, but their progeny are usually more accommodating. The peach tree did not thrive in Greece in Aristotle's time, but it now flourishes in the severe climate of North Germany. The first orange trees introduced into Italy were repeatedly cut down by the frosts, and it was only when plants were raised from seed grown in the country that the orange became established. The common domestic fowl, if transported from a temperate to a tropical climate, or, conversely, from a tropical to a temperate climate, will seldom long survive the change, although its progeny will in course of time adapt themselves to either climate.

These facts seem to point to some power in the reproductive system which modifies the organism only at birth.

Are organic variations purposive, or are they mere aimless deviations from the original type? He would be a rash man who would affirm that anything in nature was aimless or purposeless. We may not be able to discover what Nature's purpose is any more than we may know the means through which she operates, but it is as certain that every action of Nature has an aim as that every effect has a cause. Darwin speaks of variability as "spontaneous" and "accidental," leading one to believe that he held the opinion that it was subject to no law, but he explains that he uses these peculiar terms to express that it is brought about by "unrecognized or unassignable" causes. Nor is he less vague in describing the conditions under which variations arise. All that he contends for is that organisms should have "a capacity for change," or "a tendency to ordinary variability," which is a somewhat odd explanation.* On the other hand, Dr. Wallace boldly asserts that variations are fortuitous. He no doubt sees clearly enough that if certain variations arise under certain conditions, and proceed in definite lines, there would be no case for the plaintiff, and natural selection would be out of court; he therefore insists that variations are indefinite, or that

* *Origin of Species*, p. 210.

they break out "in every direction"*—we presume, like
the bristles on the hide of a hedgehog or buds on the
stalks of Brussels sprouts. †

It is evident, however, that if variations were mere
accidental or "sportive" occurrences (to use one of
Darwin's expressions), and the organism had no hand in
shaping its own destiny, it would inevitably become a
mere agglutination of parts or organs and modifications
of organs, without symmetry and without harmonious
action. Darwin admits that an organism is affected by
the conditions of existence; but he fails to state what
action is taken by the organism under such circumstances.
Now, the organism does take action; it responds to the
conditions when these affect it, whether favourably or
unfavourably, by adapting itself to these conditions. In
every instance in which there is a change in the conditions of the life of an organism there is not only a
struggle to live, as Darwin maintains, but (and this is
the main point) there is a struggle to adapt or to
modify itself to the new conditions. According to the
Darwinian theory there is no struggle in the latter sense
at all, for the variations are supposed to occur quite
promiscuously, not to say accidentally, and the favourable
ones are few in number and slight in degree. ‡   But an

---

* *Natural Selection*, p. 290.   † See *ante*, p. 24.

‡ Dr. Wallace, in his latest work, admits the force of the objection that has been often raised against the theory of slight variations—namely, that they would be practically useless to the

occasional variation, even if favourable, and more especially if a slight one, would stand a poor chance of surviving, as it would almost inevitably be lost by intercrossing in the course of a few generations. On the other hand, if the conditions were uniform and prolonged, and the tendency of the organism to adapt itself to these conditions were persistent, there would be a constant succession of favourable variations, from which new races and new species might be evolved ; but without a constant succession of variations of the same kind the formation of new species would be an impossibility.

We have said it is not to be supposed that the external conditions have any direct influence in producing variations, or that variations can arise independently of the organism. The directing power in every instance must proceed from the organism itself, the external conditions being only the occasion for, not the cause of, the variations. The organism is the active agent of its own transformation. It cannot defy the laws of its environment, but it can accommodate itself to them. A tree growing in a sheltered spot will send its stem straight up into the air, a tree in an exposed situation will expand horizontally instead of vertically,

individuals possessing them ; and he has brought forward a mass of interesting facts which show that variations are not slight, but that the various parts and organs vary from 5 per cent. to 25 per cent. under or over the average.—*Darwinism*, chap. iii.

and a leaning tree will throw up buttresses from its roots
in order to enable it to withstand the force of gravitation.
Not only the organism as a whole, but every part of it,
possesses this power of adaptation. Thus, as Mr.
Henslow points out, if a large fruit, like an apple or a
pear, hang vertically it grows symmetrically, but if the
pedicle projects obliquely from the branch it thickens
along the upper side, forming a sort of buttress running
down the stalk, which also tends to thicken.* John
Hunter fed a sea gull (*Larus tridactylus*) on grain for a
year, and succeeded in so completely hardening the inner
coat of its stomach, which is naturally soft and adapted
to a fish diet, that in structure it resembled the horny
skin of the gizzard of a pigeon. Dr. Edmonstone informs
us that another species of gull, *Larus argentalus*, of the
Shetland Islands, changes the structure of its stomach
twice every year, according to its food, which consists of
grain one half the year and fish the other half. These
cases prove that the stomach of a carnivorous bird may
be transformed into that of a grain-eater ; and Dr.
Holmgren's experiments with pigeons show that the
converse is equally true, and that if a pigeon be fed on
meat for a sufficiently long period its gizzard can be converted into a carnivorous stomach. It is this power of
self-adaptation that is the key to organic modification.

The cuticle of the rower's hand, the muscles of the
blacksmith's arm, are both enlarged by use ; but, although

* *Floral Structures*, p. 623.

the enlargement follows from the friction and the strain, the latter is not the cause of the enlargement. If the hand or the arm were inanimate bodies the friction and the strain would produce erosion, whereas in living bodies we see the effect is precisely opposite. The law of Use and Disuse is a mechanical law to which the organism has to conform, and it conforms by strengthening the parts subjected to wear and strain. What is it that the law of Use and Disuse teaches us? The wing-bones of the domestic duck weigh less than those of the wild species, the result of disuse; the diminished snout of the domestic pig is the consequence of the diminished use of that organ; the drooping ears peculiar to our domestic animals are due to the disuse of the muscles of the ear ; the muscles of the wings of the homing pigeon are stronger than those of the bard or the pouter, because the former have been subjected to greater strain than the latter. For the same reason birds inhabiting islands where there are no beasts of prey lose the use of their wings; the eyes of the mole are rudimentary in size ; animals which inhabit dark caves are wholly or partially blind, as are ento-parasites. Nor can we exclude man from the operation of this law. From the time when quaternary man crouched in a cavern and tore the raw flesh from the bones with his teeth, how much of the improvement which he has since achieved is due to his own unaided intelligent efforts ? In all those cases the organisms have been the active agents in the modification of their own structures with-

out the aid of any form of natural selection. They
have not only exhibited "a capacity for change" but a
capacity for change in a definite direction. These facts
are familiar enough, yet their full significance has not
been sufficiently appreciated. The terms use and disuse
hardly convey all that the facts express. Use and disuse
simply mean function and its absence, but the facts imply
more than that ; they indicate intelligence, purposiveness,
and effort. The terms Effort and Abstinence would better
designate the law in the matter.

Organic changes are the result of physiological—not
mechanical—causes. The external conditions only affect
the organism by way of stimulating it to action, but
there must be a response on the part of the organism
before any organic change can take place. The nature
of the response will depend on the nature of the stimulus ;
if the latter is favourable the organism will react by
unfolding itself to its utmost capacity, and if in a
vigorous condition it may develop latent capabilities
which might to us appear to be new variations ; if, on
the other hand, the influence is adverse, the organism will
react in a different manner. If it cannot overcome an
adverse influence it will endeavour to adapt itself to it.
A plant will react upon sunshine and rain, which are
favourable conditions, by a vigorous growth ; when exposed to strong biting winds, which are adverse, it will
expand horizontally, and assume a stunted form of
growth. In the one case it will unfold, and in the other

case it will modify itself. In the latter instance the
modification will not be the effect of natural selection
but of what we should rather call Self Adaptation. The
external conditions affect the organism, and the organism
effects the modification.

Customs, usages, and language come into existence
and survive or disappear much in the same manner as
organic modifications. They are adapted to meet certain
conditions of society, and they endure or perish with the
conditions which call them forth. Thus a language, like
an organism, is variable and perishable; it grows, decays,
and dies ; it has its rudimentary and abortive stages, and
its fossil remains ; it is subject to constant changes—words
in common use to-day become obsolete to-morrow, old
words take new meanings, and old meanings find expression in new words. As new conditions of life evolve
new variations in structure, so new events, new dis-
coveries, and new habits call forth, but do not create,
new phrases. And, to pursue the analogy further, it
would not be one whit less absurd to say that language
was the result of natural selection than that organic
modification was due to the same cause.

Summary.—I have attempted to show that there are
two factors concerned in organic modifications, the external and the internal, or the nature of the conditions
and the nature of the organism, and that the one factor
can have no effect without the co-operation of the other.

The external conditions, as food, climate, situation, have great influence in modifying organisms, but only by way of inciting or stimulating them to action. The conditions of life can of themselves effect no organic change; in every instance there must be a response on the part of the organism before any modification can take place. These conditions, if uniform, pronounced, and prolonged, will, according to their nature, invariably incite the organism to change in a definite direction. If the conditions are favourable, the organism will respond by expanding itself to its utmost capacity ; if unfavourable, it will respond by adapting or adjusting itself to its new conditions. The modifying action will, in every instance, proceed from the organism, not from the external conditions, which are merely the occasion, not the cause, of the modifications. I have also indicated that it is not only the organism as a whole, the personality as it were, that possesses this power of modification, but every part of it, so that no organ or part of an organ is deprived of this power of self-adjustment. The law of Use and Disuse, or, as I prefer to call it, of Effort and Abstinence, prevails throughout the organic world, modifying each organ or part of an organ as circumstances require, and always in a manner calculated to enable the organism to maintain life and to carry on its functions to the best possible advantage.

CHAPTER VII.

WHEN one discovers what he imagines to be a law of nature, he is apt to believe that he has arrived at the limit of knowledge on the subject. This, however, is a mistake. A law throws no light on the origin of phenomena. A law is only a relation—a regular concurrence of phenomena, the precedence and sequence of certain events in relation to other events. Thus we often speak of the law of use and disuse, as if the mere statement of the law was a sufficient explanation of the phenomena. That the use of an organ, for instance, is followed by certain results, only explains the fact of the concurrence, not the cause of the phenomena. Having ascertained the law of Adaptation, we shall now proceed to inquire by what process adaptation is brought about. As it is not a mechanical process it would be vain for us to attempt to explain it, as Darwinists do, on mechanical principles, and therefore natural selection will be of no assistance to us here. That it is a physiological process admits of no doubt ; but this admission brings us no nearer to the origin of the phenomena in question. We say that external conditions influence the organism, and that the organism responds ; but how is it that the organism comes to respond ? Reflex action will not account for the phenomena, for we have still to account for reflex action.

Are the actions of organisms guided by intelligence, or are they the result of mechanism ? If the former, then such actions will be purposive, like those of human beings, and will be subject to the same mental laws. If the latter, we shall have to ask, Whence the mechanism ? As we cannot suppose it has come by accident, it must have been designed ; if designed we shall have next to ask who is the designing agent ; and as we cannot speak of a self-acting mechanism we shall still have to ask what sets the mechanism in motion. Although Darwin, as we have seen, often makes use of such terms as "contrivance" and "purpose" when speaking of organic phenomena, he says not a word that would lead one to suspect that he believed the phenomena in question to be a manifestation of purposiveness or intelligence, either on the part of the organism or on the part of some *ab extra* power acting through it. We cannot, however, draw a line arbitrarily at man, or at mammals, or at the higher vertebrata, and say, below this there is only reflex action or automatism. On the other hand, if we admit that animals possess intelligence we cannot logically deny the same attribute to vegetative organisms, as many of the higher orders of plants exhibit greater intelligence than many of the lower animals.

Some plants indeed display wonderful intelligence in their movements. Look, for instance, at what takes place in fertilization. As soon as the pollen is mature, the stamens move towards the pistils, or the pistils towards the stamens, or both towards each other. These movements never commence till the pollen is mature, and cease the moment fertilization has been accomplished; and what is still more extraordinary, if it should happen that fertilization has already been brought about by artificial means or by the visits of insects, these movements, being then unnecessary, never take place. Fertilization is impossible if the pollen should by any means become wet, hence plants take the greatest possible care to prevent this occurring. Many plants close their corolla when it is about to rain, or when the air is moist with dew; others hide their flowers under their leaves at night. Even aquatic plants have to keep their pollen dry—almost an impossible thing for them, one would imagine, yet they contrive to accomplish it. The water-nut (*Trapa natans*), for instance, lives under the water till flowering time, when the petioles become filled with air, which raises the flower buds to the surface, when florescence and fructification take place. Immediately this has been accomplished the petioles discharge the air, which is replaced by water, and the plant sinks again to the bottom. All this may be put down to mechanism; but we may just as well ascribe human action to mechanism, as man has a far more complicated structure than is possessed by the poor water-nut or by any known species of plants.

We find abundant evidence of intelligence in the lowest forms of animal life. A rizopod, for instance, is to all appearance, though not actually, a formless speck of albuminous matter, without even a cell wall. It lives in water, and it puts forth processes of its body (pseudo-podia) to suck the chlorophyll out of minute plants, or to seize its prey, for some species are omnivorous, by enclosing it with its processes. Yet this apparently formless microscopic animal leads an independent life, and performs all the functions of a complex organism. It has no special organs of prehension, no mouth, no stomach, no intestinal canal, and yet it can seize, swallow, digest and assimilate its food ; it has no organs of loco-motion, and yet it can move about at will ; it has no sense organs, yet it can distinguish light from darkness, and, judging from the manner in which it selects its food, it can discriminate aliment from what is not aliment. Referring to the amœba, which is also a mere jelly speck, Dr. Carpenter confesses that we can "scarcely conceive that a creature of such simplicity should possess any distinct consciousness of its needs," nevertheless goes on to say :—

Suppose a human mason to be put down by the side of a pile of stones of various shapes and sizes, and be told to build a dome of these, smooth on both surfaces, without using more than the least possible quantity of a very tenacious but very costly cement in holding the stones together. If he accomplished this well, he would receive credit for great intelligence and skill—yet this is exactly what these little 'jelly specks' do on a most minute scale, the 'test' they construct, when highly magnified, bearing comparison with the most skilful masonry of man. From *the same sandy bottom*, one species picks up the *coarser* quartz grains, cements them together with *phosphate of iron* secreted from its own substance, and thus constructs a flask-shaped 'test' having a short neck and a single large orifice. Another picks up the *finest* grains, and puts them together with the same cement into perfectly spherical 'tests' of the most extraordinary finish, perforated with numerous small pores, disposed at pretty regular intervals. Another selects the *minutest* sand grains and the terminal portions of sponge spicules, and works these up together, apparently with no cement at all, by the mere 'laying' of the spicules, into perfect white spheres, like homœopathic globules, each having a single fissured orifice. And another, which makes a straight many-chambered 'test,' that resembles in form the chambered shell of an orthoceratile—the conical mouth of each chamber projecting into the cavity of the next—while forming the walls of its chambers of ordinary sand grains rather loosely held together, shapes the conical mouths of the successive chambers by firmly cementing together grains of *ferruginous* quartz, which it must have picked out from the general mass. To give these actions the vague designation 'instinctive,' does not in the least help us to account for them; since what we want is to discover the mechanism by which they are worked out; and it is most difficult to conceive how so artificial a selection can be made by a creature so simple.*

The arcella, another marine animal, is a minute speck of protoplasm with a shell in which are fine perforations through which it protrudes itself by pseudopodia, like the

---

* *Principles of Mental Physiology*, 4th ed., p. 42.

rizopod. This animal, which also leads an independent
life, has the extraordinary power of forming cavities in
its protoplasm into which air is admitted or withdrawn,
whereby it can increase or reduce its specific gravity, and
so lower or raise itself in the water at pleasure. The air
bubbles are seen to be continually changing their size and
position when the animal is in motion, showing that they
are regulated by a will within.*

There is evidence that the semi-independent cells which
go to make up a complex organism—what we call organic
cells—are not destitute of intelligence. A complex
organism may be said to be a community of cells. Every
organic structure consists exclusively of cells and the
products of cell action—that is to say, of tissue. A
complex organism is developed out of a germ cell, which
propagates by fision, the divided cells forming new cells,
which again divide and form other cells, and so on till a
sufficient variety and number are produced to form the
several kinds and qualities of tissue required for its construction. The cells are the sole agents employed in this
work. When we see a colony of bees constructing their
combs on a concerted plan, we say that they exhibit intelligence; why should we deny to organic cells the like
attribute, more especially as they exhibit far more constructive power than the insects? When we see an
orderly system of government carried on in any community we naturally conclude that its members are

* Pflüger's *Archiv für Physiologie*, vol. ii.

intelligent beings. There is ample evidence of such an orderly system of government in organic communities. They have a complete system of division of labour. Every kind of tissue has its special cells for its production; and besides these constructive cells there are others which are set apart for more special work—representative cells we may call them, as they receive and convey impressions from and to all parts of the organism and regulate all its functions. These are the nerve cells. We find them distributed throughout the organism in groups and masses—some presiding over local organs; some in large masses in important centres and lines of communication; some in still larger masses, as in the brain, which exercises correspondingly large functions. Each group is subordinate to the centre immediately above it, but is at the same time capable of determining and maintaining certain movements or functions of its own without the intervention of the supreme centre.*

In the case of the constructive cells, when any injury happens to a complex organism of which they form a part, they immediately proceed to repair it. When a bone which has been fractured does not unite, they will absorb the smaller portions and round off the two ends; if the ends unite but overlap they remove the projecting parts, thus showing that they possess the faculty of discrimination. Supposing a breach occurred in the embankment

* Maudsley's *Physiology of Mind*, p. 109.

of a canal or a river, and we saw hundreds of men at work shovelling earth into the gap made by the overflow of the water, we should say that these men showed purposiveness and intelligence. The microscope reveals to us the fact that organic cells co-operate for similar purposes. When a fision takes place in any part of an organic structure, masses of cells congregate around the spot, and industriously set themselves to work to close the wound. Why should we refuse to acknowledge intelligence in this case while accepting it in the other ? *

Again, when any foreign body gets lodged in any part of a living organism, the local cells at once set themselves to work in order to expel it. This they do by a process of suppuration. If any secretions accumulate from any cause in the interior of the organism, and there is no natural outlet for these to escape, the cells at once proceed to open a new passage, and the remarkable thing is that such passages remain open only so long as is necessary for the expulsion of the secretions, when they are immediately closed.

The cells have also the power of restoring lost organs. Crabs, lizards, tortoises, salamanders, and other cold-blooded animals, and many species of annelids and insects, have this power of reparation. If a worm be cut in two, it will reproduce a tail on the one section and a head on the other, and, according to Spallanzani, if divided into three parts the centre portion, which is deprived of

---

* *See* Appendix B.

both head and tail, will reproduce both, but the head first, as if the cells were conscious that this was the more important of the two. In certain fishes the restoration of fins is effected in the order of their importance, the caudal taking the precedence, then the pectoral and ventral, and lastly the dorsal fins. Voit records the case of a pigeon which had its brain restored after five months' deprivation; after the fifth month a white mass began to show itself where the hemispheres had been removed. This white mass possessed the appearance and consistency of brain substance; it united with the peduncles which had not been removed, and the animal gradually recovered its intelligence. Blumenbach mentions an instance where an eye had been restored within a year from the time of deprivation, the optic nerve not having been injured, and the complete restoration of the crystalline lens has often been observed in animals from whom it had been removed.*

Mechanism, chemical affinity, or reflex action throws no light upon the subject, as these always behave in a definite manner, according to their nature; but the process of restoration varies according to circumstances, and the cells turn aside from their ordinary functions to perform work of a totally different kind, which they perform according to the exigencies of each particular case. In every instance they allow for complicated or disturbing relations. "If a limb be amputated half-way up

* See Appendix B.

the thigh, for instance, the cells near the surface of the
stump work together by dividing and developing in
various ways, so as to reproduce the limb. Supposing
now that the line of amputation is oblique or irregular,
this does not affect the result. The cells concerned in
the process allow for the irregularity. If, for example,
two-thirds of the extensors of the knee, half the thigh
bones, and a third of the hamstring muscles, are amputated, the missing part of each of these structures is
nevertheless reproduced in its proper proportion." *   If
actions of this kind do not exhibit intelligence, what does ?

That a simple cell should have this constructive
capacity is certainly extraordinary; but it is not more
so than that a lowly organized cold-blooded animal
should have the power of restoring lost organs, while
the more highly organized warm-blooded animals do
not possess it, or only to a very slight extent ; while
this again is not more wonderful than the action of the
ganglion cells which preside over the viscera.   The
ganglia, which are dispersed through the muscles of
the heart, the stomach, the intestinal canal, and other
organs, control the actions of those organs altogether
independent of the Ego.   So far, indeed, from the Ego
having any direction of these organs, it is not even
conscious of their action or even of their existence.   In
this case there is complete subdivision of labour : the Ego
takes the general control of the external movements

* J. S. Haldane in *Mind*, vol. ix., p. 30.

of the organism, while the cells are the active agents of all internal movements and of all organic modifications. The bearing of all this on organic modifications will now be obvious. The cell is the source of all change. As a modifying agent, its power exceeds that of the complex organism of which it forms a part, just as cold-blooded animals have reparative power which the higher warm-blooded animals do not possess. In growth and decay, in the increase and decrease in size, and in complexity of structure, the cell is the controlling agent. A slight pressure or friction, causing irritation, or a stimulus on a particular cell or a group of cells, will often produce the most unexpected results. An insect alighting on a stamen or pistil of a flower will, as we have seen, often modify the functions of the sexual organs, and change the whole system of reproduction. The irritation caused by the visits of insects will also, according to Kerner, produce hairs or sticky glands on the stems or flower stalks of certain plants, while the browsing of bushes by mammals will induce the growth of prickles and thorns, and only on those portions of the plants which are within reach. Pressure on the cells of any part of a plant will lead to a special formation of that part; friction on the skin will result in an increase in the number of epithelial cells, and a thickening and hardening of the epidermis ; excessive heat, acting on the epidermic cells, converts the wool on the sheep's back into hair ; light, acting on the chlorophyll cells of plants and the pigment cells of animals, produces changes in the colour of the foliage and flowers of plants, and of the wool, hair, and fur of animals.

Lamarck's theory of organic evolution differs materially from that here presented. According to Lamark, it is the wants and efforts of the organism which effect changes in its structure.* Unfortunately, the experience of mankind shows that no want, however severely felt, no effort, however protracted, ever produced a new organ or modified an old one, and we have no reason to believe that the experience of other organisms differ from our own in this respect. On the other hand we have seen how skilfully the cells manipulate a fractured bone, how they close wounds, get rid of internal secretions, and even restore lost organs, so that it does not require a very great stretch of imagination to suppose that they also possess the power of adapting themselves to new conditions of life.

The cell is the biological unit. It is the irreducible vital entity; it is the seat of life and energy; it is the

---

* "The production of a new organ in an animal results from the supervention of a new want *(besoin)* continuing to make itself felt, and a new movement which this want gives birth to and encourages;" and he illustrates this law by the case of the gastero-pod mollusc finding " the need of touching the bodies in front of it, makes efforts to touch these bodies with some of the foremost parts of its head," hence the flow of nervous and other liquids to these parts and the extension of the nerves abutting on them.

key that unlocks the mystery of organic modifications. And here we may observe that if the cell be the biological unit, Life cannot be a synthesis, or the sum of all the activities of the organism, as many physiologists main-tain. Lewes defines Life to be "the expression of the whole organism." In a certain sense it is so, as the whole organism is animated by the cells which permeate its structure; and so it may be said that a multicellular organism exhibits more life than an unicellular organism. A monad can move about, but its movements would be much expedited if it were provided with special organs of locomotion. But, on the other hand, special organs would not confer upon the monad the power of locomotion, as the creature already possesses it, but only afford greater facilities for movement. Instead, therefore, of Life being the expression of the organism, the organism is rather the expression of Life, or of the vital force inherent in the germ cell. Lewes scoffs at the idea of a vital force or principle; "why not have a crystal principle," he asks, "to personify the conditions of crystallization?" Curiously enough, this is precisely what we already have, the term Affinity being called upon to do duty here, as Gravitation in physics. Both terms are used to personify the causes of certain phenomena which are incapable of more definite explanation. The fact of the matter is that Lewes has misunderstood the whole ques-tion in dispute by not bearing in mind the twofold signification of the term Life. He regarded Life as an

effect ; but Life is also a cause. The phenomena exhibited by a living animal may be called the life of that
animal ; but all phenomena must have a cause, and the
cause of vital phenomena we, in the absence of more
definite knowledge on the subject, call vital force.   To
say that the cause of life of a complex organism is the
organism itself, would be to confound cause with effect,
as life already existed in the germ cell out of which the
organism was formed.   He also objects that the vital
principle is a special force called up for the occasion.   I
reply, so are the phenomena also special.   A living
organism is unlike any other known body.   It grows,
decays, dies ; it increases in size not by the addition of
matter to the outside, like crystals, but by converting
matter into its own substance ; it is distinguished by its
tendency to cyclical changes, whereas inert matter is in a
state of stability and repose ; it has the power of reproduction ; it can transmit to its progeny its own characteristics to the minutest shade, and it is capable of adapting
itself to adverse conditions of existence.   These phenomena are peculiar to living bodies, and are sufficiently
marked to distinguish them from all phenomena of
whatever kind.

The cell is also the psychological unit.   It is the irreducible element which feels, thinks, and wills.   The
term *Mind*, like the term Life, is also liable to be misunderstood, as it has a twofold signification.   It is used
to denote—(1) what Descartes called " the thinking sub-

stance," or the soul, and (2) the product of that substance or soul, namely, thought.* In the former sense
Mind is the psychological unit, in the latter it is the
product of that unit. The psychological unit is a marvel
and a mystery. It possesses potentialities beyond anything in nature. What a wonderful body it fashions for
itself; what vigilance it exercises in maintaining this
body in health and vigour, in repairing injuries, in
adapting it to new conditions of life! Look at the
process which goes on from the birth of the germ-cell till
its development into a full-grown and complex organism :
how the ovum becomes a cluster of cells ; how these send
out processes which, in a few hours, days, or weeks, take
the form of a spinal column, a head, legs, arms, wings, or
fins, according to the type of animal to be formed, till
the whole complicated organism is gradually constructed.
And this germ-cell has within it the accumulated experience of untold generations. It inherits and trans-

* The term *Mind* is also used in a third sense, as denoting a
something that is not matter, as immaterial or non-extended ; but
now when force is identified with matter this meaning is very
properly going out of use. It is obvious that we cannot determine what mind is unless we have first ascertained what matter is.
We can only speak of things relatively, as causes or as effects.
Moreover, we cannot speak of mind, which is projected in space,
as non-extended. Mind may be a property of matter, or matter
may be a property of mind, just as force may be a property of
matter, or matter a property of force ; but to speak of a thing as
not-matter or immaterial, when we do not know what matter is,
seems somewhat absurd.

11

mits the smallest physical and psychological peculiarities. Quite extraordinary, for instance, is the inheritance and transmission of mental maladies. These maladies may lie dormant for generations, and yet they may break out in the organism at the same period of life, and actually in the same manner, in successive generations as they have done in some remote ancestor. Lucas mentions a case where all the members of a family in Hamburg, distinguished through four generations for great talents, went mad at the age of forty. Esquirol gives an instance where father, son, and grandson committed suicide at the age of fifty; and Voltaire knew a man whose father and brother died by their own hands at the same age as the man himself, and in precisely the same manner.

Many courageous attempts have been made from time to time to explain mind in terms of matter, but all such attempts assume the existence of a sensitive organism, which is surely the thing that has first to be accounted for. It has been contended, for instance, that the original element of all psychical experience is a nervous shock; that it is by the differentiation and combination of nervous shocks that sensation in its simplest form arises and that by the agglutination and agglomeration of simple sensations are ultimately produced all the complex phenomena of mind. If by sensation is here understood a physical event, then we have not yet reached the simplest mental state; if, on the other hand, sensation is a mental event, how has the physical nerve-shock been converted into a mental sensation ?

Sensation is again supposed to be a something intermediate between matter and mind (with a great deal of the former in it), out of which mind is, by some unexplained process, developed. " Sensation is so conspicuously a physiological process," says Lewes, "that many writers exclude it from the domain of mind."* J. S. Mill says—" The immediate antecedent " of sensation "is a state of body ;" † Mansell, that " it is not an affection of the mind alone, but of an animated organism —that is, mind and matter united."‡ No doubt sensation is an affection of the organism, but so also are the emotions, and so are perception and volition. The term sensation conveys the idea of passivity, but it includes far more than this. From its nature it must include consciousness ; indeed, consciousness is its very essence. Sensation must be the property of a conscious subject. When I say " I feel," I assert that this is my mental condition ; I discriminate between what is self and what is not self ; between what is feeling and what is not feeling. Thus the very thing we profess to evolve out of sensation— namely, mental activity—already exists. Even granting that sensation is an intermediate something, and that mental phenomena are evolved out of it, we have still to explain how we come to have sensation at all, so that the

---

* *Physical Basis of Mind*, p. 335.   † *Logic*, ii., p. 436.
‡ *Metaphysics*, p. 92.

question of the origin of mental phenomena remains unsolved.

Nor can it be maintained that mental phenomena are the result of organization. No doubt organization is necessary for the fuller development of the phenomena in question, just as a high-class instrument is necessary for the accomplishment of high-class work of any kind ; but as work of a certain kind can be done with very inferior tools, or with no tools at all, so can mental operations be carried on without a highly specialized organism. The steam engine, to which a living organism has been compared, has no function unless its mechanism be perfect ; but psychical force can operate without a brain or a nervous system. A highly specialized organism exhibits a greater amount of intelligence than a microscopic cell, just as we have seen a multicellular organism shows more vitality than an unicellular organism ; but here the question is not one of quantity but of kind. The production of mental, any more than of physical phenomena, is not the exclusive function of large masses. The atom as well as the mountain, the smallest satellite as well as the largest planet, has its attractive force. If the phenomena of mind are not the product of the germ cell, at which period in the history of the organism do they come upon the scene ? Mental phenomena are by many supposed to be the product of organization ; but organization is an effect rather than a cause. If organization were the

cause of mental phenomena we should have to ask, What
is the cause of the phenomena of organization ? Organization, however, is no more necessary for the performance of mental functions than are special organs for
the purpose of locomotion. The hydra are sensitive
to light although they have no eyes ; many zoophytes
have no ears, yet they contract themselves when vibra-
tions of sound are propagated through the medium
in which they live ; polypes, which have neither
nerves nor brain, feel and move without a nervous
system. Every physiologist is familiar with cases in
which the loss of one sense-organ is followed by the
increase or intensity of the other sense-organs which
have been left intact. Many blind and deaf people
acquire such a wonderful delicacy of touch as to be able
to dispense with the organs of sight and hearing, of
which Abercrombie gives numerous instances. Laura
Bridgman, with only the tactile sense left her, but which
was very acute, and the sense of smell very imperfectly
developed, could nevertheless acquire impressions of
objects which are ordinarily obtained only through the
sense-organs she was deprived of ; and Kruse, who was
completely deaf, had a bodily feeling of music. Modern
science has demonstrated that Democritus was correct
when he said that " all the senses are but modifications
of touch."

But we may be asked if mental phenomena are not
evolved from a nerve-shock, are not developed out of

sensation, and are not the result of organization, how have they originated ? We have anticipated the answer to this question : they are the product of the psychical unit. We cannot get behind the psychical unit, any more than we can get behind the biological unit, or the physical unit or atom. We may reduce the cell to its chemical elements, but by no combination of these elements can we produce the phenomena in question.

I have attempted to draw an analogy between Mind and Life, and I have assumed them to be two independent agencies. But why this dualism ? Are there really two agents of this mysterious sort, one producing mental phenomena and the other biological phenomena ? The functions of Life and of Mind are so much alike that it is often difficult to distinguish between them. Maudsley admits that mental processes closely resemble vital processes, but, like most physiologists, he assumes that the former are the result of the latter.* Johannes Müller, defines Life to be " a rational creative force," which is equivalent to saying that the phenomena of life are due to a psychical cause. The same agent often produces both mental and organic phenomena. When I will to move my arm, the willing is psychical, the movement is organic. It may be proper enough to describe the one set of phenomena as subjective or mental, and the other as objective or organic, but we are not justified in ascribing to them different origins, the one set to a

* *Physiology of Mind*, p. 43.

psychical and the other to a vital force. The idealist will have no hesitation in determining which should have the precedence.

But how, it may be asked, can the many become one, unity arise out of diversity, the permanent Ego out of an aggregate of conscious, unstable units, without some unifying agent? Nothing seems so certain as my personality, and yet nothing is so evanescent. In the first place, let me remark, the Ego is a synthesis; it dissolves itself into a series of sensations and memories as soon as we attempt to grasp it. How often has one to ask oneself in the course of his life, Can I really be the same individual I formerly was, when my views of life, my aspirations, and my sentiments generally are now so utterly unlike what they were in times past? It is only by pulling oneself together by a strong effort of memory, that one can believe in his own past continuity. What ideas or sentiments has a man of mature age in common with those of his youth? Our mental being has been described as a series of states of consciousness—a constantly flowing stream, not a stagnant pool—and this is what we should expect to find from the nature of our organization. The grouping and massing of nerve cells, with their connections ramifying through all parts of the organism, permit of a succession of sensations to be transmitted to the supreme centre or brain. This organ we may suppose to be the seat of the Ego, a view which receives confirmation from the fact, which has been

demonstrated by Ferrier, Fritch, Munk, and others, that
the organs of sense, which are indispensable to the Ego,
are represented in the brain and nowhere else.*

In the second place, there need be no incoherence in an
aggregation of units organically united. The human
brain is composed of two distinct hemispheres, but though
divided we have not a divided consciousness, at any
rate in our normal state. The social community is
made up of independent units, yet we personify this
community when we speak of Public Opinion and the
National Conscience. Here, however, as we have said,
the units are independent; but in a complex organism
they are united by nerve fibres, and are at the same
time in such close proximity in the brain that they
appear as if they were fused together. It has been often
observed that when two individuals have been long and
intimately associated together they unsconsciously adopt
each other's mode of looking at things; in the case of
twins there is often an extraordinary mental and physical
similarity between the two, extending in some instances
to identity and simultaneity of action. Ribot says:—

It has been observed that when there is perfect physical
similarity between twins, which is not rare, it is always accompanied with moral similarity. Moreau saw at Bicêtre two young
men who were so much alike that one would be taken for the
other. They both possess the same monomania, the same dominant
ideas, the same hallucinations of hearing; they never speak to any
one, nor do they communicate with one another. An exceedingly

* See Appendix C.

curious fact, often observed by the attendants and by myself, is this : from time to time, at irregular intervals of two, three, or more months, without appreciable cause, and by the entirely spontaneous action of their malady, a very marked change occurs in both brothers at the same period; often on the same day they quit their habitual state of stupor and prostration, and earnestly entreated the physician to give them their freedom. I have seen this repeated even when the two brothers were separated from one another by a distance of several miles.*

Moreover, as cells are reproduced by fision of the parent cell, which divides itself into two parts, both of which continue to live, being, according to Weismann, practically immortal,† as there is thus a continuity in the life of the organic cells there is also a continuity or identity in the organism. On the physical side we find the organic cells, nerve cells, and nerve fibres distributed throughout the body, and in connection with these a mass of minute nerve cells and nerve fibres in the brain ; on the mental side, we have a corresponding diffusion and concentration of psychical units more or less intimately associated. But the facts adduced do not lead up to the conclusion that the concentration of psychical units, which we may call the Ego, and the diffusion of psychical units, which we

* *Heredity*, p. 266.

† The process of fision in the amœba has been recently much discussed, and I am well aware that the life of the individual is generally believed to come to an end with the division which gives rise to two new individuals, as if death and reproduction were the same thing. But the process cannot truly be called death. Where is the dead body ? What is it that dies ? Nothing dies. —Weismann's *Heredity*, p. 25 (English trans.)

may call the Soul, are identical, although they are undoubtedly closely connected. In one sense, the Ego is the Soul, inasmuch as it represents the Soul, being in direct communication with each of the psychical units of which the latter is composed ; in another sense the Ego, being only a part of the Soul, or a synthesis of psychical units, cannot be the Soul itself. All roads lead to Rome, and all mandates are issued from there, and therefore we say Rome is the empire ; but Rome is not the empire, but only a part of it, although a very important part.

The fact that the brain alone is in direct communication with the organs of sense has important bearing on the question of the evolution of the Ego. We know self only when we are able to distinguish between the subjective and the objective, and the objective we know only through the medium of the senses. We have no reason to believe that the Ego exists in the embryo, although the latter possesses an elaborate nervous system long before birth. Even after birth the Ego of the infant is non-existent, although the evolution of its brain has far advanced. For weeks after birth the infant seems to have no definite idea of self, and even when it begins to prattle it speaks of itself in the third person. With the continuous growth of the organism increased mental capacity will follow in the ordinary course, by the multiplication, differentiation, and combination of the psychical units ; the slow process by which the co-ordination of sense impressions is brought about corresponds with the gradual development of the Ego after birth ; but the turning point is reached when the power of distinguishing between what is subjective and what is objective, between what is self and what is non-self, has been acquired.

But, although the Ego has its own mode of being, it does not lead an independent existence. Its relation to the Soul is that of the stem to the root. That part of a plant that is above ground leads a very different life from that which is below. Some species of plants have even separate reproductive systems, the roots producing tubers and the flowers seed. The part above ground can only live in the sunshine and by breathing the free oxygen of the atmosphere, while the part below supplies it with moisture and nourishment. In like manner the Ego lives its own conscious life, has its own sphere of action, its own pleasures and pains, its own hopes and fears, its own sentiments and aspirations ; but it nevertheless leads a dependent life, for, as by a process of capillary attraction the stem and branches draw their aliment from the root, so the Ego draws its instincts, its intuitions, and its inspirations from the Soul. The Ego has supreme control of the external relations of the organism, and is entirely unimpeded in its movements ; the Soul presides over the internal relations, maintains the functions of the organism, every part of which it pervades, and every part of which it modifies when modification becomes necessary.

Summary.—We have in this chapter endeavoured to
trace, as far as our ignorance will permit, organic modifications to their source. This we found to be the organic
cells. These apparently formless and semi-independent
organisms exhibit intelligence, and have the power of
self-modification and recuperation to an extent possessed
by no other living being. They are the sole agents employed in the construction, and afterwards in the maintenance, of the most complex organisms, and their
economic and social organization is both comprehensive
and complete. When an injury occurs to any part of
the organism they collect in force on the spot for the
purpose of effecting repairs, which they execute with
singular skill and judgment, varying the means employed
according to the circumstances of each particular case.
We have seen wherein this view differs from that put
forward by Lamarck, who ascribes modifications to the
wants and efforts of the organism as a whole, and
not to the biological units in the parts affected. As we
have seen that the germ-cell exhibits the phenomena we
call mind, it cannot be said that mind is the result of
organization ; we should say that organization is rather
the result than the cause of mental activity. There is,
first, the action of the environment, and, secondly, the
reaction of the organism. This action is either stimulative or irritative ; if stimulative, the organism reacts
by expanding itself, and this often results in the development of latent qualities ; if irritative, the organism

reacts by modifying those parts which are injuriously affected. But the system of division of labour requires that local wants should be provided for by the local cells, so that when the action of the environment (which is purely mechanical) adversely affects an organ, or any part of an organ, the modification which follows would not be brought about by the Ego (which has only a dim idea of something being wrong somewhere, and knows nothing of the changes that may be necessary or how to effect them), but would be due to the local cells, whose special function it is to repair, replace, and to modify the parts immediately under their control. On the physical side we have the germ-cell, which, by a process of division, becomes a community of organic and nerve cells, a mass of the latter forming the brain; on the mental side we have a corresponding distribution of psychical units, and a concentration of a proportion of such units forming the Ego. The community of psychical units within the organism we may call the Soul; a synthesis of these psychical units we may call the Ego, and the functions of these two are so distinct that the latter not only does not direct, but is not even conscious of, the operations of the former.

APPENDIX.

A.—MIMICRY.

Note to page 29.

THERE is not only a general resemblance between animals and their environment, but there is sometimes also a special likeness (often greatly exaggerated), between certain animals, especially insects, and some particular objects to which they may resort in order to escape observation, such as a flower, a piece of moss, a green or withered leaf, or even the droppings of birds. Many insects are reddish on red soil or on red sandstone, speckled on granite, white on chalk or limestone, sometimes even shaded to suit the colour of a particular stone wall or a small patch of gravel, while other varieties of the same species, only a few yards off, will be coloured to match some other kind of background. One species of insect may also be found associated with another species of a totally different character, but to which it may have a general resemblance, as, for instance, an edible caterpillar with a kind which is non-edible, by which means the former may often escape from the depredations of insectivorous birds, which avoid non-edible species. In such cases the edible species is supposed to imitate or mimic the likeness of the non-edible. Can we imagine such resemblances to be brought about by the action of natural selection? Why should we have recourse to a process involving, as this does, the sacrifice of innumerable species, and requiring for its operation almost infinite time, when the same result might be brought about at once, and without any sacrifice whatever? It is only amongst insects whose habits are sluggish, and who have no means of defence that these extraordinary resemblances occur. Many such animals when they find themselves helpless will feign death. Touch a woodlouse, for instance, and it will roll itself up into a ball and remain perfectly still. The common speckled weevil rolls off the leaf it is sitting on at the approach of a possible enemy, and drawing in its legs and antennæ, it will thus become undistinguishable from the pellets of earth among which it falls. A butterfly will attempt to escape capture by alighting on some object to which it has a general resemblance, and will thereupon remain motionless till the danger is past. Butterflies are protectively coloured only when at rest; when in flight their slow fluttering motion renders them very conspicuous; hence, as every collector knows, when chased they suddenly disappear by alighting on some object coloured like themselves, whereby they escape observation. To account for these likenesses to special objects, animate or inanimate, we have only to assume that these defenceless insects have intelligence enough to perceive that their safety lies in escaping observation. That such animals possess colour-perception is evident from the fact that they invariably select an appropriately coloured background when pursued by an enemy; and that they make their selection with the view of escaping observation is manifest from the confidence they exhibit by remaining motionless even when an enemy approaches within a few inches of them. There is a wide difference between this kind of selection, however, and that with which the name of Darwin is identified. In both cases a certain amount of colour variation is assumed; but in the one case the organism selects an environment to suit itself, in the other the organism is modified to suit the environment.

B.—THE CELL.

Note to page 138.

SINCE the discovery of the cellular theory by Schleiden and Schwann it has become more and more apparent that the cell is a factor of supreme importance in physiological investigation. Virchow even goes the length of saying that if we have to choose between the (multicellular) organism and the (individual) cell that we should prefer the latter to the former. Why should we hesitate to acknowledge the significance of the cell in the province of psychology also ?

" All life is bound to the cell, and the cell is not merely the vessel of life, but it is itself the living part. . . . What is the organism ? A community of living cells, a little state, well provided with all appurtenances of upper and under officials, servants and masters, great and small."—Virchow's *Vier Reden*, p. 55.

" The whole protoplasmic system must be conceived as an individualized organism—*i.e.*, a living, moving, proper being, consisting of nucleus, peripheral envelope, and radial or net-like uniting members, and found within its self-formed shell, the cellulose wall, in continual motion, which consists in a gliding hither and thither, and a consequent shifting and constant remodelling of the internal articulation. As the mollusc not only constructs its own shell, but moves within the same, so the protoplasmic body within its cellular membrane. Not the currents in the bands, not the cell nucleus, not the primordial sac *per se*, are the seat and cause of the movement. The whole protoplasmic body, which is not a substance but an organism, moves in all its parts, now simultaneously, now alternately, as indivisible, amoebiform, vitalized, proper being, which of course in the higher plants is only partial existence of a larger whole."—Hanstein's *Botanische Ztg.*, 1872, Nos. 2 and 3.

" So certain and intimate is the sympathy between the individual nerve-cells in that well-organized commonwealth which the nervous system represents, that a local disturbance is soon felt more or less distinctly throughout the whole state. When any serious degeneration of the ganglionic cells of the cord exists, there is not only an indisposition or inability to carry out as subordinate agents the commands which come from above, but there is a complaint sent upwards; a moan of discontent or pain reaches the supreme authority. That is the meaning of the feelings of weariness, heaviness, achings of the limbs, and utter lassitude which accompany disorder of the spinal centres; and the convulsive spasms and the local contractions or paralysis of muscles are the first signs of a coming rebellion. If the warnings do not receive timely heed, a riot may easily become a rebellion; for when organic processes, which normally go on without consciousness, force themselves into consciousness, it is the certain mark of a vital degeneration. If the appeal is heard in vain, then further degeneration ensues. Not only is there irregular revolutionary action of a subordinate, but there is *pro tanto* a weakening of the supreme authority; it is less able to control what is more difficult of control. When due subordination of parts exists, and the individual cell conforms to the laws of the system, then the authority of the head is strengthened. A foolish despot, forgetting in the pride of his power that the strength and worth of a government flow from and rest upon the well-being of the governed, may fancy that he can safely disregard the cry of the suffering and the oppressed; but when he closes his ears to complaints, he closes his eyes to consequences, and finally wakes up to find his power slipped from him, and himself entered upon the way of destruction. So it is with the nervous system: the cells are the individuals, and as in the state, so here, there are individuals of higher dignity and of lower dignity; but the well-being and power of the higher individuals are entirely dependent upon the well-being and contentment of the humbler workers in the spinal cord, which do so great a part of the daily work of life. The form of government is that of a constitutional monarchy, in which every interest is duly represented through adequate channels, and in which, consequently, there is a proper subordination as well as co-ordination of parts."—Maudsley's *Physiology of Mind*, p. 180.

12

The action of the amœboid cells (leucocytes) in inflammation is thus graphically described by Dr. Sutton :—

"If we summarize the story of inflammation, as we read it zoologically, it should be likened to a battle. The leucocytes are the defending army, their roads and lines of communications the blood-vessels. Every composite organism maintains a certain proportion of leucocytes, as representing its standing army. When the body is invaded by bacilli, bacteria, micrococci, chemical or other irritants, information of the aggression is telegraphed by means of the vaso-motor nerves, and leucocytes rush to the attack : reinforcements and recruits are quickly formed to increase the standing army sometimes twenty, thirty, or forty times the normal standard. In the conflict cells die, and often are eaten by their companions ; frequently the slaughter is so great that the tissue becomes burdened by the dead bodies of the soldiers in the form of pus, the activity of the cell being testified by the fact that its protoplasm often contains bacilli, &c., in various stages of destruction. These dead cells, like the corpses of soldiers who fall in battle, later become hurtful to the organism they in their lifetime were anxious to protect from harm, for they are fertile sources of septicæmia and pyæmia—the pestilence and scourge so much dreaded by operative surgeons. The analogy may seem to many a little romantic, but it appears to me to be warranted by the facts which have been placed before the reader." —*General Pathology*, p. 127.

C.—PERSONAL IDENTITY.

Note to page 152.

THIS theory of the Ego may help to explain certain abnormal phenomena, such as hallucinations, as when one person supposes he is another person ; and double consciousness, of which there are many known instances. Professor Janet has astonished the scientific world by his recent investigations, which go to show that there are sometimes as many as three distinct states of consciousness, or three different personalities, in the same subject. He made observations on several hysteric patients in the Havre Hospital who exhibited this extraordinary peculiarity. The case of Léonie, a peasant woman of the ordinary French type, and 45 years of age at the date of observation, is particularly interesting. We give Professor Janet's own remarks about this singular case :—

When in her normal state this poor peasant woman is a serious and rather sad person, calm and slow, very mild with everyone, and extremely timid. To look at her one would never suspect the personage which she contains. But hardly is she put to sleep hypnotically than a metamorphosis occurs. Her face is no longer the same. She keeps her eyes closed, it is true, but the acuteness of her other senses supplies their place ; she is gay, noisy, restless, sometimes insupportably so ; she remains good-natured, but has acquired a singular tendency to irony and sharp jesting. Nothing is more curious than to hear her after a sitting when she has received a visit from strangers who wished to see her asleep. She gives a word portrait of them, apes their manners, pretends to know their little ridiculous aspects and passions, and for each invents a romance. To this character must be added the possession of an enormous number of recollections, whose existence she does not even suspect when awake, for her amnesia is then complete. . . . She refuses the name of Léonie, and takes that of Léontine (Léonie 2), to which her first magnetizers had accustomed her. " That good woman is not myself," she says ; " she is too stupid." To herself Léontine (or Léonie 2) she attributes all the sensations and all the actions—in a word, all the conscious experiences—which she has undergone in somnambulism, and knits them together to make the history of her already long life. To Léonie 1, on the other hand, she exclusively ascribes the events lived through in waking hours. I was at first struck by an important exception to the rule, and was disposed to think that there might be something arbitrary in this partition of her recollections. In the normal state Léonie has a husband and children, but Léonie 2, the somnambulist, while acknowledging the children as her own, attributes the husband to " the other." This choice was, perhaps, explicable, but it followed no rule. It was not till later that I learned that her magnetizers in early days, as audacious as certain hypnotizers of recent date, had somnambulized her for her first accouchements, and that she had lapsed into that state spontaneously in the later ones. Léonie 2 was thus quite right in ascribing to herself the children, since it was she who had had them, and the rule that her first trance state forms a different personality was not broken. But it is the same with her second state of trance. When, after the renewed passes, syncope, &c., she reaches the condition which I have called Léonie 3, she is another person still. Serious and grave, instead of being a restless child, she speaks slowly and moves but little. Again she separates herself from the waking Leonie 1. "A good but rather stupid woman," she says, "and not me." And she also separates herself from Léonie 2. "How can you see anything of me in that crazy creature," she says. "Fortunately I am nothing for her."*

It is a singular feature of the case that Léonie 1 knows only of herself; Léonie 2 knows of herself and of Léonie 1; Léonie 3 knows of herself and also of both the others. Professor Janet also discovered that these different personalities not only appeared in succession but actually existed simultaneously. Thus he found when Léonie 1 was fully absorbed in conversation with a visitor that Léonie 2 would hear his voice when he addressed her in a whisper, and would respond by obeying his orders or by gestures or by writing, Léonie 1 meanwhile going on with her conversation quite unconscious of what Léonie 2 was doing as it were behind her back. Professor Janet explains these phenomena by supposing that the unifying or co-ordinating power of the ordinary personality was unable to grasp all the facts of consciousness (hysterical subjects being often extremely *distraites*, and unable to attend to more than one thing at a time), and that the *residuum* which it could not synthesize formed other personalities. Be this as it may, there can be no doubt that a plurality of consciousness, whether successive or simultaneous, is capable of explanation on the theory we have presented, as any alteration in the grouping of the psychical units would involve a change of personality, and if the total possible consciousness were split up into parts these parts would arrange themselves into separate personalities, which might successively or simultaneously assert themselves.

* *De L'Automatisme Psychologique.*

George Robertson and Co., Little Flinders-street, Melbourne.

www.ingramcontent.com/pod-product-compliance
Lightning Source LLC
Chambersburg PA
CBHW021808190326
41518CB00007B/495

* 9 7 8 3 7 4 2 8 6 5 0 0 7 *